CLOCK GUIDE

Identification With Prices

by

Robert W. Miller

Published by

Wallace-Homestead Book Company
1912 Grand Avenue
Des Moines, Iowa 50309

i

ISBN 0-87069-360-3

Photography by
Dudenbostel and Thurman
Knoxville, Tennessee

Printed in the United States of America by
Wallace-Homestead Co.
Des Moines, Iowa 50309

Clock-Wise

Here we are with the Seventh printing of our ever-popular CLOCK GUIDE, Identification With Prices. Ours is the **only** CLOCK GUIDE that gives you two prices — a high and a low — because clock prices vary from coast-to-coast, at auctions and antiques shows, from swap meets to flea markets. You're assured of knowing the facts when you use our CLOCK GUIDE!

When you compare our CLOCK GUIDE, using the only clear photographs, with some of those other so-called price guides that use tired, worn out catalog reprints, compare and you'll see why our CLOCK GUIDE continues to be the very best.

As the cost of food, fuel and clothing continue to sprial ever upward, use common sense when you refer to our CLOCK GUIDE. The prices quoted here are those generally paid for clocks in good-to-excellent condition. The same holds true in our CLOCK GUIDE No. 2.

Every effort has been made to eliminate errors, but the author and the publisher cannot be held responsible for mistakes, either in typography or judgement as to any prices in this book.

The theft of quality clocks continues to be big business in the United States. Learn how to mark your clocks with some form of identification and never talk to strangers until they have properly identified themselves.

Thanks for asking, but under no circumstances will we identify the owners of the clocks used in this CLOCK GUIDE and CLOCK GUIDE No. 2.

Robert W. Miller

Contents

The TIME of Your Life

Probably the most important word in the world today is "time." We get up, go to bed, eat, arrive, leave, work, play and, unfortunately, even fight wars by it. Through the ages we've struggled to master this complex, this elusive element—time!

Our cultural and social environment can be traced back some 6,000 years to the first sundial. In the beginning, people worked by day and slept by night, the sun and the moon dictating their hours of labor or sleep. It took a long time to discover that a complete revolution of the earth on its axis took 23 hours, 56 minutes, and 4.9 seconds. Thus, the 24-hour day.

Dividing a year into months and days is a man-made invention, resulting in the calendar. To date no calendar has been found to be absolutely perfect. The Babylonian calendar divided the year into 12 parts named after the 12 signs of the zodiac. The early Mayan and Aztec tribes worked out a similar system, except they used 18 months of 20 days each. Our present-day calendar is based on the old Roman calendar.

Thousands of years ago the Chinese told time by burning a piece of rope, slowly, from knot to knot. Later, candles marked with notches were used. The Greeks and Romans allowed water to drip from a hole punched in the bottom of a container. None of the methods mentioned were really reliable, but it was the best they had.

The first mechanical clock was constructed by Henry De Vick for King Charles of France, about 1360. Its basic parts were power, transmission, escapement and hour hand. Almost 300 years passed before the smooth, even swing of the pendulum was brought into use. Galileo is credited by most with first proposing it, but a Dutch clockmaker, Christian Huygens, is believed to be the first to use the pendulum to control the rate of motion of a weight-driven clock.

Finally, an Englishman, Robert Hooke, invented the universal joint and the wheel-cutting engine. At long last, accurate clocks became possible. Before the year 1700 pendulum clocks were available for home use, with minute hands on the dial, and shortly after with second hands.

The deadbeat escapement for regulator clocks was invented in 1715, and six years later, Graham's mercurial pendulum came into being. Six more years brought Harrison's gridiron pendulum. Both Thomas Tompion (1639-1713), England's leading clockmaker, and his partner, Graham, were buried in Westminster Abbey—a high honor in those days to be bestowed upon a clockmaker.

For hundreds of years clocks were an expensive luxury. They were handmade, built one at a time, and always costly. Sometimes when the maker died, he took his particular secrets with him to the grave.

Two things changed all this. Mass production, beginning in Europe and reaching its height in America, lowered the cost of clocks. The competition to develop a completely accurate timepiece was fierce. In the 18th century the British government offered a prize of 20,000 pounds to the inventor of an accurate chronometer, a seagoing timepiece necessary to finding a ship's position by navigation. John Harrison, a self-taught horologist, won the coveted prize. There are different versions as to exactly how long it took him to collect.

It was only natural, when the early colonists migrated to America, that they would bring their skills, including the art of clockmaking. "Handmade" wasn't just a word—it was an accepted way of life in those days. As early as 1650, the first tower clock in America was in use in Boston. When George Washington was inaugurated at the Old Federal Hall in Wall Street, New York City, a public clock, built in 1716 by Joseph Phillips, tolled out the news to the viewers.

Probably the first man to build a complete tall clock in America was Abel Cottey who came to Philadelphia on the same ship with William Penn in 1682. Encouraged by Benjamin Franklin, Edward Duffield made Philadelphia's first town clock around 1750.

Benjamin Banneker and Peter Hill were two distinguished Negro clockmakers. Banneker plied his trade in Baltimore, and Hill made clocks in New Jersey, shortly after the American Revolution. At Lancaster, Pennsylvania, Anna Marie Leroy, daughter of a Swiss clockmaker, made a complete brass clock movement as early as 1750. In 1802, Hannah Montanden carried on her husband's clock business for six years after his death. And Maria Nicollet had her own shop in Philadelphia in the late 1700s. Some say she was the first woman watchmaker in the United States.

Some confusion has been created by clocks signed "Made by Ethel North, Wolcotville, Connecticut." Ethel was a man.

Fine tall clocks dominated the American market, finally giving way to the mass-produced Connecticut shelf clock. This state gained fame, and rightfully so, as a pioneer on the American industrial scene. A cabinetmaker and boat builder, Ebenezer Parmele, gave the Nutmeg State its first tower clock, erecting it in a church steeple in his hometown of Guilford in 1726.

The Cheneys of East Hartford made beautiful clocks, and two of their apprentices went on to worldwide fame, Benjamin Willard and Eli Terry. If any one state contributed more to the making of fine clocks than any other, it most certainly would have to be Connecticut. Among the distinguished clockmakers were Thomas Harland, Norwich; Daniel Burnap, East Windsor; Gideon Roberts, Bristol; James Harrison, Waterbury. Bristol, Plymouth, and Waterbury were the heart of clockmaking in New England.

The age of "one clock at a time" had passed. When mass-produced clocks became the vogue, some method of distribution had to be found. First on horseback, then by wagon, the clockmakers either took to the road or hired others to sell their clocks for them. It would not be unfair to state that these Yankee peddlers paved the way for the American industrial system.

An astounding fact is that as early as 1807 Eli Terry contracted with a firm in Waterbury, Connecticut, to make 4,000 "hang-up" wood clock movements at four dollars each! In 1814, Terry manufactured his first wooden movement 30-hour Pillar and Scroll clock that sold for 15 dollars! President Madison personally signed one of Terry's patents in 1816.

Joseph Ives, Bristol, Connecticut, is credited with making the first metal movements with mass production machinery. In the 1820s, the largest clock producers were Chauncey Jerome (in partnership with his brother Noble) and Elijah Darrow, operating as "Jerome and Darrow."

In 1837, when America was in the midst of a business depression, the clockmakers were in sore need of something new to attract buyers to their product. Chauncey Jerome came to the rescue by manufacturing a one-day weight-driven clock. The clock was named "OG" after the style of molding used, but was popularly called Jerome's "clock for everyone." It later sold for one dollar with Jerome's own company leading the way by selling more than 500,000 in a single year!

Of all the clockmakers none enjoyed more success than Seth Thomas. His first plant was at Plymouth Hollow, Connecticut, in 1813. The town, renamed Thomaston in 1866, has been making Seth Thomas clocks for 158 years.

Around 1850, springs replaced weights for motive power and after 1860 Thomas' company began to make calendar clocks. Until his death in 1859 Seth Thomas took an active part in the operation of his business. No clockmaker in history contributed more to the industry than this great man.

Elias Ingraham (1805-1885); Elisha Manross (1792-1856); James Birge (1785-1862); Irenus Atkins (1792-1882); Solomon Spring (1826-1906); Elisha N. Welch (1809-1887)—all made major contributions to the clock industry in America.

No history of clocks would be complete without mentioning the Willard brothers. Benjamin, Ephraim, Simon, and Aaron were all masters in their own right. Simon was perhaps the most important because of his development of first the one-day wall clock; then the "Massachusetts shelf" model, followed later by the eight-day "coffin clock," and finally his masterpiece, the "banjo" clock. He probably invented also the forerunner of the automobile speedometer, the odometer. He called his invention a "perambulator." It measured the distance a carriage had traveled.

In closing, let's clear up an old wives' tale about why the hands on a display clock are usually at 8:18. Some say it was the exact time President Lincoln was shot. This is wrong by more than an hour and a half as Mr. Lincoln was shot after 10 p.m. and died the next morning at 7:23 a.m. Probably closer to the truth is that when the clock hands are at 8:18 the maker has the maximum room, without crowding, for his name or other advertising material.

If you enjoy modern living, say a silent "Thanks!" to the early clockmakers who gave you "the TIME of your life."

The Care and Maintenance of Old Clocks

Don't try to take apart or repair an old clock unless you are very sure you know what you're doing. The next few pages may save you from ruining a rare clock or, more important, from ruining yourself. Having a clock's mainspring explode in your face is comparable to jumping feet first into a bathtub full of razor blades.

My best advice is to take your "sick" clock to a reliable clock repairman. The information given here is not intended to encourage you to start tinkering with old clocks, but to give you some idea of the complexity of the repair job. There's good reason why you should approach it with caution. On the other hand, there are things you can do yourself in the way of cleaning and restoration.

A lot of clock repairmen are personal friends of mine and I've spent many hours discussing this segment of the *Wallace-* *Homestead Clock Guide* with them. I don't blame them for smiling silently when I pursue certain aspects of clock repair. The repairing of old clocks isn't something that one learns overnight. In all cases, my friends came up with the answer: It's fun to collect clocks but it can get very demoralizing when the amateur ends up with a table covered with foreign looking parts.

Experience—that's the key word. This is **not** to say you can't learn how to repair and maintain your own clocks. This **is** to say, however, that you should know what you're doing before you start!

Here are just a couple of problems:

Knowing how many coils are held under tension in the spring barrel is important because the spring must be rewound the exact number of turns it's unwound or the "stop work" gets fouled up. The stop work is a safety device to prevent injury to the escape wheel from overwinding, or to prevent undue force coming on the pendulum by jamming the weights. It is also a means to utilize only the middle portion of the long and powerful spring.

Clocks have a greater number of teeth notches than watches, determined by the number of turns desired for the arbor. When the clock has run down 16 turns of the barrel, the fingers will again meet on the opposite side, so the barrel will be allowed to turn backward and forward 16 revolutions, being stopped by the fingers at each extreme.

If that statement isn't enough to make you leave your clocks alone, read on.

"Motion work" is the name given to the wheels and pinions used to make the hour hand go once around the dial while the minute hand goes around 12 times. The formula is confusing to the inexperienced and is only mentioned here because, without it, you'll never get your clock in proper running order. The relation of the wheels and pinions is 8 to 1 and 7.5 to 1; $7.5 \times 8 = 60$.

In the motion works, it's 3 to 1 and 4 to 1; $3 \times 4 = 12$. As I said, it's confusing.

Now for some things that you can do:

A clock, like an automobile, needs lubrication from time to time. Don't drown it in oil, however. A light oiling once or twice a year is sufficient. The easiest way to oil your clock is to buy a hypodermic syringe at your local drug store. Don't use a heavy lubricant. Sewing machine oil is permissible, though it's a good idea to ask a clock repairman the type of oil he uses. The hypo needle makes it easier to get at the hard-to-reach gears.

In restoring the exterior of old clocks, remember that a little dirt is better than a lot of scrubbing of gilt work or clock dials. Age cracks add to the value of a clock. When you repaint a dial, you lower the value and you destroy the originality of the piece.

Now, let's talk about refinishing the wood case. It may be walnut, mahogany, mahogany veneer, cherry, or oak, the last being the most popular in the lower-priced models. A particular prize is a case made of chestnut. To the untrained eye, the case may look worn or, in general, "beat up."

Before you decide to remove the works and refinish the case, take these steps:

If the case has a glass tablet or glass door, remove the glass.

Be sure to save the paper label if there is one in the case, on the back, or on the bottom. These labels identify the clock and should be preserved at all costs. If torn or slightly mutilated, carefully paste the pieces of the label down with regular paper glue. Then put on a coat or two of clear varnish to assure its preservation. If you have to refinish the case, mask off the label before you start scraping or using paint remover.

Don't refinish unless necessary. Try first to remove the dirt and age stains with mild soap and water. Gently! Don't ever use soap and/or water on gold gilt parts. Denatured alcohol will remove most fly spots and other dirt from the gilt without harming it. Again, gently! After you've "washed" the case, let it dry, then rub gently with any good paste or liquid wax. The result may be better than refinishing.

If you must finish and if the case is solid wood, you can use paint remover. But if it's veneer, and a great many of the early cases were, go slow. Reglue any loose pieces of veneer, using a toothpick to slip the glue under the loose pieces. Scotch tape will hold down the veneer until the glue dries.

Don't get over-anxious with your sanding. Fine steel wool or extra fine sandpaper is all you'll need. When you're ready to replace the original finish, use a mixture of clear varnish and naphtha, 50-50, for the first coat. Steel wool lightly between all coats. The second and third coats can be applied directly from the can. When the finish is completely dry, rub it gently with a lint-free cloth, then wax.

When you find a "collectible" clock with glass tablets or dials missing, or so badly damaged they have to be replaced, check with a clock repairman. He may be able to order the proper tablet or dial at a nominal cost. On the other hand, many tablets and dials can be salvaged by an artist who knows his business.

Old clocks are turning up almost every day in attics, basements, antique shops and at auctions. You're just as likely to find a "gem" as the next collector. If you get one, talk to a clock authority before you venture forth into the mysterious world of clock repair.

The repairing of old clocks is fast becoming a lost art. Someday in the not-too-distant future, we may find it difficult, if not impossible, to locate someone who can properly repair and maintain a clock collection.

Until that day arrives, if you feel a clock is worth buying, respect it to the extent that you give it tender loving (and professional) care.

Good hunting!

1. American-made, 8-day, lever escapement clock, mahogany case, time and strike, **$45-$60.**

2. American-made, miniature steeple, time and alarm, **$65-$80.**

3. European, miniature Grandfather clock, time and strike, **$95-$120.**

4. Lux Clock Mfg. Co., Waterbury, Conn., miniature shelf clock, time only, **$50-$60.**

5. Ansonia Clock Co., Ansonia, Conn., "Balloon" clock, 8-day, time and strike, **$240-$265.**

6. Jerome & Co., Bristol, Conn., "Flying pendulum" clock, sometimes called "Ignatz," 1884-1885, listed in New Haven Clock Co. catalog, **$140-$155.** (reproduction shown).

7. Seth Thomas, Thomaston, Conn., long alarm, metal case, 1-day, **$65-85.**

8. American-made shelf clock, miniature lever escapement, time and strike, **$90-$115.**

9. American-made novelty clock, in shape of banjo, time only, **$90-$120.**

10. French "Balloon" clock, 8-day, lever escapement, time and strike, **$125-$145.**

11. American-made miniature pendulum clock, Pat. July 23, 1878, time and strike, **$65-$80.**

12. German musical, time and alarm, **$80-$95.**

13. Terry Clock Co., Pittsfield, Mass., shelf clock with pendulum, time and alarm, **$85-$110.**

1. Sandoz-Wuille, Swiss, 8-day car clock, **$48-$60.**

2. Western Clock Co., La Salle, Ill., Westclox ironclad alarm clock, **$28-$40.**

3. Esberger Bros., Jeweler, Cincinnati, alarm clock made for their customers, **$29-$39.**

4. Western Clock Co., Westclox miniature alarm, **$20-$28.**

5. New Haven Clock Co., New Haven, Conn., double figure type clock, 8-day, gold plated, time and strike, **$115-$140.**

6. American-made, 30-hour, lever escapement, metal cased clock, time and strike, **$68-$82.**

7. Ansonia Clock Co., Brooklyn, N. Y., round top table clock, time and strike, **$98-$122.**

8. New Haven Clock Co., wood base, brass case, time and intermittent alarm, **$55-$62.**

9. Waltham Watch & Clock Co., Waltham, Mass., 8-day car clock, **$55-$65.**

10. National Watch Co., Elgin, Ill., 8-day car clock, **$47-$56.**

11. American-made, metal cased clock, time and alarm, **$35-$45.**

12. Lux Clock Mfg. Co., Waterbury, Conn., "Cupid" novelty clock, time and strike, **$92-$122.**

13. Western Clock Co., Westclox, "Big Ben" alarm, **$24-$36.**

14. Western Clock Co., Westclox, "Baby Ben" alarm, **$24-$36.**

15. Seth Thomas, Thomaston, Conn., brass lever for locomotives, 8-day, Pat. April 16, 1878, time only, **$350-$425.**

1. Waterbury Clock Co., Waterbury, Conn., shelf clock, walnut case, 8-day, time and strike, **$130-$160.**

2. French double figure mantel clock, brass/marble, time and strike, **$295-$365.**

3. Waterbury Clock Co., oak kitchen clock with barometer and thermometer, time and strike, **$210-$265.**

4. Waterbury Clock Co., carriage clock, 8-day, strike and repeat, **$195-$265.**

5. Ithaca Calendar Clock Co., No. 10 Farmer's model, walnut, perpetual calendar, time and strike, **$610-$725.**

6. New Haven Clock Co., single figure mantel clock, cast metal case, black wood base, open escapement, time and strike, **$325-$425.**

7. English watchman's clock, punch-type, fusee movement, walnut case, **$255-$295.**

1. Waterbury Clock Co., Waterbury, Conn., shelf clock, open pendulum, time and strike, **$88-$105.**

2. Waterbury Clock Co., walnut shelf clock, time and strike, **$155-$190.**

3. United Electric Co., Brooklyn, N.Y., FDR—The Man of the Hour clock, plain dial, 30-hour, time and alarm, **$92-$135.**

4. F. W. Jansen, Chicago, Ill., "Nitelite" alarm clock, **$48-$60.**

5. Waterbury Clock Co., walnut shelf clock, mercury pendulum, 8-day, time and strike, **$210-$245.**

6. Waterbury Clock Co., walnut shelf clock, 8-day, time and strike, **$170-$195.**

7. American-made iron front shelf clock, possibly Seth Thomas, **$140-$165.**

8. American-made electric table clock, carved wood case, 1920s, time only, **$32-$44.**

1. German shelf clock, time and strike, **$110-$152.**

2. Waterbury Clock Co., miniature schoolhouse clock, time only, **$153-$185.**

3. Elisha Hotchkiss, Jr., Burlington, Conn., column mantel clock, wooden movement, 30-hour, weight driven, time and strike, **$295-$365.**

4. Chelsea Clock Co., Chelsea, Mass., wardroom clock, "U.S. Marine Corps" on dial, time only, **$170-$210.**

5. Seth Thomas electric chime clock, **$72-$92.**

6. Nicholas Muller's Sons, New York City, flat top black marble mantel clock, outside escapement, time and strike, **$120-$145.**

7. Ithaca Calendar Clock Co., No. 9 shelf cottage, walnut, perpetual calendar, 8-day, time and strike, **$1,400-$1,850.**

8. German miniature carriage clock, time and strike, **$95-$125.**

1. Sangamo Corp., Springfield, Ill. (Subsidiary jointly owned by Hamilton Watch Co. and Sangamo Electric Co.), ''Sangamo'' electric clock, **$142-$165.**

2. Dutch, porcelain hanging wall clock, time only, **$38-$50.**

3. Birge, Mallory & Co., Bristol, Conn., shelf clock, 8-day, roller pinion, weight driven, time and strike, **$425-$525.**

4. American-made sterling silver bedside clock, 8-day, time and strike, **$42-$60.**

5. Ansonia Clock Co., Brooklyn, N. Y., marble mantel clock, matching candle stands on either side, **$260-$285.**

6. Waterbury Clock Co., Waterbury, Conn., walnut kitchen clock, time, strike and alarm, **$160-$185.**

1. Mission wall clock, oak case, time only, **$72-$92.**

2. International Time Recording Co. (later became part of IBM), time clock, **$155-$195.**

3. American-made kitchen wall clock, 8-day, time only, **$35-$45.**

4. Ansonia Clock Co., regulator wall clock, time and strike, mahogany veneer, **$310-$345.**

5. German mantel clock, time and strike, **$110-$145.**

6. German 3-piece mantel set, bisque, time only, **$152-$175.**

1. Seth Thomas, oak wall clock, Regulator No. 1, weight driven, time only, **$720-$825.**

2. Galusha Maranville, Winsted, Conn., octagon drop wall calendar clock, rosewood, time and strike, Pat. March 5, 1861, **$625-$750.**

3. Brewster & Ingraham, Bristol, Conn., walnut Gallery clock, time only, **$245-$295.**

4. New Haven Clock Co., chime mantel clock, **$120-$165.**

5. "Jerome & Co." (trade name used by New Haven Clock Co.), "Christmas Tree" kitchen clock, time and strike, **$195-$245.**

1. Vienna Regulator wall clock, baby 2-weight, time and strike, **$450-$575.**

2. Mission wall clock, time and strike, **$80-$95.**

3. Waterbury Clock Co., metal cased mantel clock, outside escapement, time and strike, **$120-$160.**

4. Perry & Shaw, New York City, shelf clock, wooden dial, 30-hour, weight driven, time and strike, **$195-$245.**

5. American-made, desk calendar clock with alarm, metal case, **$50-$65.**

26

1. Seth Thomas, black iron front mantel clock, twin marbleized columns on each side, time and strike, **$158-$195.**

2. German wall clock, time only, **$135-$170.**

3. Waterbury Clock Co., metal cased, bronze single figure mantel clock, time and strike, **$220-$250.**

4. American-made, Mission-type wall clock, green/white glass behind pendulum, time only, **$85-$100.**

5. New Haven Clock Co., oak kitchen clock, time, strike and alarm, **$140-$175.**

28

1. European case, American movement wall clock, walnut case, time and strike, **$195-$235.**

2. Ansonia Clock Co., single figure mantel clock, cast metal case, 8-day, time and strike, **$310-375.**

3. Sessions Clock Co., Forestville, Conn., mantel "Advertising" clock, calendar-type, originally had "Calumet Baking Co." on front, **$325-$400.**

1. E. N. Welch (Forestville), Bristol, Conn., "La Reine" desk clock, lever escapement, 1-day, brass plated, Pat. Sept. 17, Oct. 11, 1878, **$120-$165.**

2. United Electric Co., Brooklyn, N.Y., FDR—The Man of the Hour clock, animated dial, bartender's arm "shakes" drink, 30-hour, time and alarm, **$92-$135.**

3. Welch, Spring & Co., Bristol, Conn., "Lucca" shelf clock, rosewood case, 8-day, time and strike, **$255-$290.**

4. Ansonia Clock Co., Brooklyn, N. Y., "Chrystal Palace" clock, walnut base, mirrored sides, 8-day, time and strike (glass dome removed for photography purposes), **$410-$500.**

5. Wm. L. Gilbert Clock Co., Winsted, Conn., metal cased, bronzed, mercury pendulum, 8-day, time and strike, **$220-$275.**

6. Sessions Clock Co., Forestville, Conn., iron front case, brass trim, 8-day, time and strike, **$110-$145.**

7. E. N. Welch, shelf clock, mahogany veneer, painted tablet, 8-day, time and strike, **$120-$150.**

8. German "game" clock, cast metal, time only, **$48-$63.**

9. R. Lalique (French) table clock, pressed glass case, cut glass dial, battery operated, signed "R. Lalique," 1890s, **$155-$185.**

1. Waterbury Clock Co., Waterbury, Conn., Calendar No. 43, walnut, 8-day, 1860s, time and strike, **$1,100-$1,450.**

2. French desk clock, brass case with finials, 8-day, time and strike, **$195-$230.**

3. Seth Thomas, Thomaston, Conn., long alarm, metal case, 8-day, **$98-$125.**

4. New Haven Clock Co., New Haven, Conn., small Gothic (miniature steeple), mahogany veneer, 30-hour, time and alarm, **$130-$165.**

5. C. Jerome, Bristol, Conn., miniature cottage clock, mahogany veneer, time and alarm, **$145-$170.**

6. Ingraham Clock Co., Bristol, Conn., oak shelf clock, 8-day, time and strike, **$98-$135.**

1. Sessions Clock Co. (Forestville), Bristol, Conn., oak kitchen clock, "Hiawatha," 8-day, time and strike, **$160-$185.**

2. Ithaca Calendar Clock Co., Ithaca, N. Y., walnut shelf steeple calendar clock, fretwork under dials, 8-day, time and strike, 1870s, **$1,700-$1,975.**

3. French bronze single figure mantel clock, porcelain inserts in cast brass base, time and strike, **$320-$365.**

1. French carriage clock, silver case, repeater with alarm, **$180-$225.**

2. English carriage clock, brass case, time and alarm, **$195-$235.**

3. English carriage clock, brass case, time and alarm, **$220-$265.**

4. Ithaca Calendar Clock Co., Ithaca, N. Y., No. 11 octagon calendar, black walnut, 8-day, time, strike and alarm, **$1,200-$1,450.**

5. French bronze single figure mantel clock, on marble base, time and strike, **$450-$520.**

6. Seth Thomas and Sons, Thomaston, Conn., walnut shelf clock, with fretwork, time, strike and alarm, **$220-$295.**

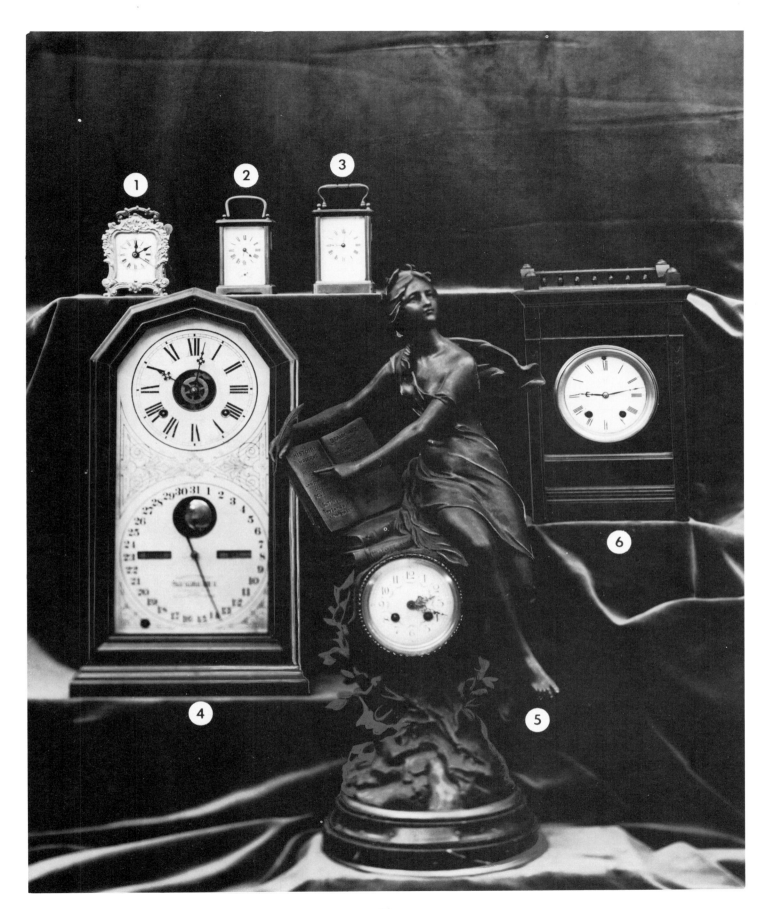

1. J. C. Brown (Forestville Mfg. Co.), Bristol, Conn., steeple fusee, mahogany veneer with painted tablet, 8-day, time and strike, **$255-$310.**

2. Ithaca Calendar Clock Co., Ithaca, N. Y., No. 3½ parlor calendar, walnut, black dials, etched glass pendulum bob, time and strike, **$3,250-$3,850.**

3. Seth Thomas Clock Co., reproduction Pillar and Scroll, walnut case, brass 8-day movement, 1929-1934, time and strike, **$135-$175.**

4. Seth Thomas Clock Co., Plymouth Hollow, Conn., rosewood case, hexagon columns, stenciled tablets, 8-day weight movement, time and strike, **$380-$480.**

5. Seth Thomas Clock Co., Plymouth Hollow, Conn., shelf clock, calendar, rosewood case, hexagon columns, stenciled tablet, 8-day weight movement, time and strike, **$1,700-$2,200.**

1. Seth Thomas Clock Co., Thomaston, Conn., column, gold leaf and gilt columns, painted tablets, Empire style, 8-day brass movement, Pat. 1867, time and strike, **$475-$575.**

2. Ithaca Calendar Clock Co., Ithaca, N. Y., No. 4½ calendar, walnut case, gold letters, time and strike, **$1,600-$1,875.**

1. Atkins Clock Co., Bristol, Conn., miniature Empire shelf clock, rosewood case, stenciled tablet, 8-day brass movement, time, strike and alarm, 1859-1879, **$360-$420.**

2. New Haven Clock Co., walnut shelf clock, 8-day, time and strike, **$145-$170.**

3. Waterbury Clock Co., Waterbury, Conn., miniature OG, mahogany veneer, gold leafed inner frame, 30-hour, time and strike, **$165-$200.**

4. Ithaca Calendar Clock Co., No. 5 Emerald, walnut case, open carving, 8-day, time and strike, **$1,900-$2,350.**

5. Ithaca Calendar Clock Co., No. 5 round top, rosewood case, 8-day, time only, **$925-$1,100.**

6. Ithaca Calendar Clock Co., small Index model, walnut case, "Index" letters in gold, made for Lynch Brothers, 8-day, time and strike, **$1,800-$2,250.**

44

1. Seth Thomas Clock Co., parlor calendar No. 5, walnut, 8-day, time and strike, **$1,150-$1,550.**

2. Gilbert L. Gilbert Clock Co., Winsted, Conn., kitchenette model, light stenciling, 8-day, time only, **$72-$92.**

3. Welch, Spring & Co., Bristol, Conn., Patti movement, rosewood case, 8-day, time and strike, 1870, **$420-$500.**

4. Ithaca Calendar Clock Co., large Index model, walnut case, ''Index'' letters in gold, made for Lynch Brothers, 8-day, time and strike, **$1,600-$1,975.**

5. Wm. L. Gilbert Clock Co., Winsted, Conn., wall, ''Eclipse,'' oak, 8-day, time, strike and alarm, **$320-$400.**

6. Seth Thomas, Fashion Model No. 2, walnut veneer, made for Southern Calendar Clock Co., St. Louis, Mo., 8-day, time and strike, **$2,400-$2,750.**

1. Seth Thomas, oak case, chimes and strikes on bells, 8-day, 1918, **$110-$160.**

2. Seth Thomas, Fashion Model No. 4, walnut case, made for Southern Calendar Clock Co., St. Louis, Mo., ''Fashion'' in gold letters on door, 8-day, time and strike, **$1,800-$2,650.**

3. Ithaca Calendar Clock Co., Bellgrade model, walnut case, wooden pendulum hangs in front of calendar movement, time and strike, **$1,200-$1,450.**

4. Seth Thomas, calendar clock, Model No. 1, mahogany veneer, 8-day, time and strike, **$585-$675.**

1. Chas. W. Fleichtinger, Sinking Spring, Pa., Victorian kitchen calendar, walnut, 8-day, time and strike, **$795-$900.**

2. Welch, Spring & Co., Bristol, Conn., wall clock, rosewood case, painted tablet, time only, **$285-$340.**

3. E. N. Welch, Bristol, Conn., walnut shelf clock, full, open columns, time and strike, **$320-$385.**

4. Seth Thomas, Standard OG, mahogany veneer, 30-hour, time and strike, **$240-$310.**

5. New Haven Clock Co., walnut shelf clock, "Etna," turned columns, time and strike, **$210-$295.**

6. Davis Clock Co., Columbus, Miss., simplified calendar, Empire case (movement made by Gilbert), 8-day, time and strike, **$610-$685.**

1. E. N. Welch Mfg. Co., Eclipse Regulator clock, long drop octagon case, with simplified calendar, oak case, time only, **$420-$525.**

2. Seth Thomas, ''Metals'' series, oak case, 8-day, time and strike, ornamented with metal, **$225-$285.**

3. Ingraham Clock Co., Bristol, Conn., mosaic, figure eight calendar clock with B. B. Lewis mechanism, rosewood case, time only, **$975-$1,200.**

4. Howard & Davis, Boston, Mass., No. 4 Regulator, weight driven, time only, ''U.S. Light House Establishment'' on front, **$2,200-$2,500.**

1. French "acorn" wall clock, with thermometer and barometer, 8-day, time and strike, **$395-$485.**

2. E. Howard & Co., Boston, Mass., No. 70 wall clock, time only, **$1,650-$2,100.**

1. Seth Thomas works, case apparently made at a later date, time and strike, **$150-$210.**

2. French wall clock, black wood with brass inner circle of painted fruit, 8-day, time and strike, **$310-$410.**

1. E. Ingraham & Co., Bristol, Conn., night and day alarm clock, **$35-$50.**

2. The Lux Clock Mfg. Co., Waterbury, Conn., Show Boat alarm clock; paddle wheel on steamboat revolves with balance wheel, animated dial, **$42-$60.**

3. Western Clock Mfg. Co., La Salle, Ill., Waralarm, made during World War II, **$32-$50.**

4. Western Clock Mfg. Co., Big Ben alarm, **$26-$36.**

5. Western Clock Mfg. Co., Intermediate Big Ben alarm clock, **$33-$50.**

6. Western Clock Mfg. Co., Waralarm clock, **$24-$40.**

7. Western Clock Mfg. Co., Westclox alarm clock, **$30-$42.**

8. Robert H. Ingersoll & Bros., Waterbury, Conn., and Trenton, N. J., alarm clock, **$28-$40.**

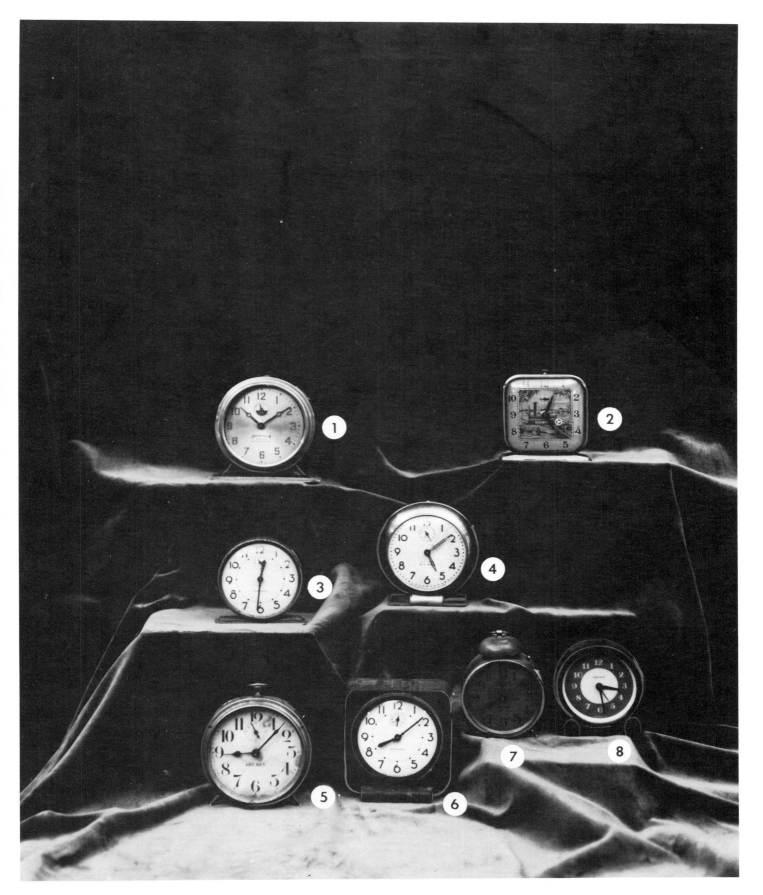

1. E. Ingraham & Co., Bristol, Conn., walnut kitchen clock, time and strike, 1890-1910, **$230-$270.**

2. L. F. & W. W. Carter, Bristol, Conn., walnut mantel clock, 8-day spring movement with B. B. Lewis patented calendar, 1862, time and strike, **$725-$840.**

3. Seth Thomas, "Eclipse" walnut kitchen clock, 8-day, time, strike and alarm, **$240-$285.**

4. Ansonia Clock Co., single figure bronze mantel clock, metal case, "Mercury," visible escapement, time and strike, **$410-$470.**

5. Jerome Mfg. Co., New Haven, Conn., mahogany miniature steeple, 30-hour time and alarm, 1845-1850s, **$125-$155.**

6. Darche Electric Clock Co., Chicago, Ill., flashlight and alarm clock, 30-hour, battery operated, Pat. 1910, **$55-$73.**

1. Welch, Spring & Co., Bristol, Conn., Patti movement, shelf clock, rosewood case, **$420-$485.**

2. J. C. Brown, Bristol, Conn., Beehive, rippled front, rosewood case, **$520-$600.**

3. Waterbury Clock Co., single bronze figure mantel clock, "Cavalier," **$495-$565.**

4. E. Ingraham & Co., Bristol, Conn., "Treasure" model, banjo, 8-day, time and strike, **$460-$555.**

5. Brewster & Ingrahams, Bristol, Conn., Twin Steeple clock, rosewood case, **$520-$585.**

6. Waterbury Clock Co., walnut kitchen clock, time, strike and alarm, **$210-$275.**

1. Ansonia Clock Co., Ansonia, Conn., and New York City, nickel novelty, "Carriage" musical alarm, Pat. March 27, 1877, **$70-$98.**

2. French, wooden case clock, **$34-$62.**

3. Wm. L. Gilbert Co., "Winlite," intermittent alarm clock, **$42-$62.**

4. Ansonia Clock Co., "Ever Ready" Plato Time Indicator, made during 1904-1906, production totaled 40,000 units, **$220-$280.**

5. English, Skeleton, 8-day fusee clock, **$300-$365.**

6. Ansonia Clock Co., mantel clock, outside escapement, **$95-$115.**

7. Seth Thomas, OG mantel clock, time and strike, **$240-$265.**

8. Waterbury Clock Co., Waterbury, Conn., novelty wall clock, porcelain case, **$80-$110.**

9. Ansonia Clock Co., mantel clock, brass case, **$225-$310.**

1. Waterbury Clock Co., Waterbury, Conn., miniature steeple, 30-hour, time only, **$160-$210.**

2. J. C. Brown, Bristol, Conn., standard type steeple, rippled front, time and strike, **$395-$445.**

3. Terry & Andrews, Bristol, Conn., shelf clock, Ansonia movement, cast iron case, mother-of-pearl inlay, 8-day, time and strike, **$260-$325.**

4. Seth Thomas, long alarm, metal case, 1-day, **$92-$115.**

5. English shelf clock, 8-day, quarter hour strike, **$215-$255.**

6. French double figure mantel clock, gold dipped, marble base, **$400-$475.**

7. French shelf clock, black marble, 8-day, 2 built-in keys, **$320-$365.**

8. German shelf clock, 8-day, walnut case, **$140-$175.**

1. Ansonia Clock Co., shelf clock, Royal Bonn porcelain case, 8-day, time and strike, **$320-$365.**

2. Seth Thomas, Empire No. 14, gold plated, mercury pendulum, 8-day, time and strike, **$460-$510.**

3. Waterbury Clock Co., miniature carriage clock, **$165-$220.**

4. Waterbury Clock Co., shelf clock, open escapement, time and strike, **$165-$220.**

5. Waterbury Clock Co., Calendar No. 36, oak, 8-day, time and strike, **$725-$925.**

6. Waterbury Clock Co., carriage clock, solid brass case, 8-day, repeater, time and strike, **$160-$210.**

7. Waterbury Clock Co., Calendar No. 42, walnut, 8-day, time and strike, **$900-$1,150.**

8. French Empire clock, open pendulum, twisted columns, time and strike, **$360-$420.**

9. E. N. Welch Mfg. Co., Bristol, Conn., octagon lever, duplex movement, 8-day travel clock, amber glass case, **$220-$275.**

1. German 400-day clock with glass dome, **$210-$255.**

2. Waterbury Clock Co., wall clock, 8-day, **$32-$42.**

3. C. and N. Jerome, Bristol, Conn., 8-day repeating, rack and snail strike, **$620-$750.**

4. French black marble mantel clock, time and strike, **$165-$210.**

5. Seth Thomas miniature clock, 8-day, **$78-$110.**

6. French double figure mantel clock, brass/marble, thread suspension, **$300-$355.**

7. Waterbury Clock Co., carriage clock with case, 8-day, **$165-$200.**

8. German, 8-day spring Cuckoo wall clock, **$165-$195.**

1. Forestville Clock Manufactory, Bristol, Conn. (one of J. C. Brown's trade names), triple decker, carved top, walnut case, painted tablets, 8-day brass movement, 1849-53, **$625-$750.**

2. Bradley & Hubbard Mfg. Co., Meriden, Conn., cast iron case, stenciling with mother-of-pearl inlay, clock movement by Ansonia, 1854-1890, **$195-$245.**

3. French shelf clock, gold gilt with porcelain dial and cap, time and strike, 1860-70, **$325-$400.**

1. E. N. Welch (Forestville), Bristol, Conn., Transition, 8-day spring movement, Empire style, mahogany veneer, gilt columns, painted tablet, **$325-$395.**

2. Ansonia Clock Co., Ansonia, Conn., and New York City, walnut shelf clock, case inlaid, brass candlestick columns, 8-day, time and strike, **$185-$225.**

3. New Haven Clock Co., Royal Bonn porcelain case, 8-day, time and strike, **$320-$400.**

4. French shelf clock, gold gilt case with columns, **$210-$265.**

5. Ansonia Clock Co., Royal Bonn porcelain case, outside escapement, **$350-$400.**

6. Ingraham Clock Co., Bristol, Conn., Venetian model, rosewood case with gilt columns, 8-day, time and strike, **$220-$265.**

1. Sessions Clock Co. (Forestville), Bristol, Conn., mahogany shelf clock, **$80-$110.**

2. American-made black iron case with white metal trim, 8-day, time and strike, **$110-$160.**

3. American-made novelty clock, metal case with columns, **$40-$52.**

4. Wm. L. Gilbert Clock Co., Winsted, Conn., alarm clock with bell on top, **$48-$62.**

5. American-made, metal case with pressed glass columns, 30-hour, **$75-$92.**

6. Wm. L. Gilbert Clock Co., miniature shelf clock, 30-hour, **$66-$78.**

7. Ansonia Clock Co., Ansonia and New York City, single figure mantel clock, ''Racine,'' gilt metal, 8-day, time and strike, **$255-$310.**

8. Seth Thomas and Sons, Thomaston, Conn., single figure mantel clock, cast iron, time and strike, **$395-$445.**

9. French shelf clock, washed gold, 8-day, time and strike, **$270-$310.**

1. French single figure mantel clock, ormolu gilt, **$295-$355.**

2. Seth Thomas, Thomaston, Conn., chronometer lever (ship's type) clock, nickel plated, 8-day, **$360-$420.**

3. Sangamo Electric Co., Springfield, Ill., Sangamo electric clock, mahogany case with 11-jewel Illinois watch movement, 1926-28, **$120-$155.**

4. Chauncey Ives, Bristol, Conn., Pillar and Scroll clock, wooden movement, mahogany veneer case, painted tablet, **$1,450-$1,750.**

5. American-made, novelty "tape measure" clock, 30-hour, **$42-$53.**

6. E. N. Welch Mfg. Co., Bristol, Conn., small cast iron "valise" clock, 30-hour, **$38-$52.**

7. Western Clock Co., La Salle, Ill., Westclox alarm, 30-hour, **$29-$39**

8. E. N. Welch, Bristol, Conn., paperweight clock, octagon lever, duplex movement, emerald green glass case, 30-hour, **$220-$265.**

9. Seth Thomas Clock Co., Fashion calendar clock, Model No. 5, long pendulum, made for Dixie Calendar Clock Co., **$2,000-$2,400.**

1. American Clock Co., Bristol, Conn., and New York City, black iron case with stenciling, 30-hour, 1849-1879, **$99-$128.**

2. French shelf clock, white alabaster with brass trim, **$170-$195.**

3. Seth Thomas Clock Co., Thomaston, Conn., miniature OG, rosewood case, 30-hour, time and strike, **$145-$165.**

4. Pratt (Daniel, Jr.) & Frost (Jonathan), Reading, Mass., Venetian style, rosewood case, 8-day, time and strike, 1832-1835, **$265-$295.**

5. E. Ingraham & Co., Bristol, Conn., Venetian style, rosewood case, stenciled tablet, 8-day, time, strike and alarm, **$168-$182.**

6. E. N. Welch Mfg. Co., Bristol, Conn., rosewood shelf clock, 8-day, time, strike and alarm **$178-$192.**

7. Seth Thomas, flat top black wood case, 3 flat columns, marble insert, each side, bronze finish trim, time, strike and alarm, **$172-$192.**

8. E. N. Welch, clock-watch, Columbus Exposition Award, 1893, **$115-$135.**

9. New Haven Clock Co., Gallery clock, black walnut, 8-day, 24-hour dial, **$255-$298.**

1. Eli Terry, Jr., Terryville, Conn., mahogany mantel clock, 30-hour, brass works, 1837, double door, **$620-$710.**

2. Seth Thomas, Plymouth Hollow, Conn., mahogany mantel clock, double door, stenciled columns, eagle splat, 8-day, wooden works, **$450-$575.**

3. Ansonia Clock Co., walnut teardrop shelf clock, time and strike, 1880s, **$365-$450.**

4. French style metal case clock, 8-day, time and strike, **$165-$225.**

5. Welch, Spring & Co., Forestville, Conn., rosewood case, round top with full pillars, 8-day, time and strike, Pat. 1868, **$195-$240.**

1. Daniel Pratt, Jr., Reading, Mass., mahogany case, 30-hour, wooden works, 1837, **$385-$445.**

2. C. & L. C. Ives, Bristol, Conn., mahogany case, triple decker, 8-day, brass movement, weight driven, 1830-1838, **$775-$900.**

3. Waterbury Clock Co., Waterbury, Conn., walnut miniature shelf timepiece, 1850-1890s, time only, **$140-$180.**

4. German musical alarm, carriage-type, **$125-$170.**

5. Ansonia Clock Co., mantel clock, bronze finish case, porcelain insert, rhinestones in bezel, 8-day, time and strike, **$300-$375.**

6. Terry Clock Co., Waterbury, Conn., single figure mantel clock, gilt metal case, 30-hour, time and alarm, 1868-69, **$210-$265.**

7. Smith & Goodrich, Bristol, Conn., miniature fusee, time and strike, 1847-52, **$560-$665.**

1. Eldridge G. Atkins, Bristol, Conn., mahogany case, 30-hour, wooden movement, 1838-1842, **$560-$665.**

2. John Birge, Bristol, Conn., 8-day weight driven, brass movement, triple decker, 1830-1860, **$725-$850.**

3. Ansonia Clock Co., bronze finish mantel clock, "Trianon," porcelain insert, 8-day, time and strike, **$260-$325.**

4. New Haven Clock Co., walnut case, mirror side shelf clock, **$295-$345.**

5. Ansonia Clock Co., single figure bronze mantel clock, "Newton," visible escapement, 8-day, time and strike, **$380-$425.**

1. Waterbury Clock Co., Calendar No. 43, oak, 8-day, time and strike, Pat. 1855-56, **$1,250-$1,550.**

2. Clark, Gilbert & Brown, New York, N. Y., OG, brass weight movement, 8-day, 1840s, ''United States Hotel'' on tablet in door, **$450-$550.**

3. Seth Thomas, round top shelf clock, 8-day, time and strike, ST hands (''S'' on 1 hand; ''T'' on other hand), **$195-$260.**

4. Ansonia Clock Co., single figure mantel clock, 8-day, time and strike, **$425-$495.**

5. Seth Thomas, miniature cottage, 30-hour with built-in alarm, ST hands, **$225-$285.**

1. Seth Thomas, Fashion Model No. 4, made for Southern Calendar Clock Co., St. Louis, Mo., Pat. March 18, 1879. (A color photo of clock No. 2, page 48), **$1,700-$2,000.**

2. Jason R. Rawson, Saxton River, Vermont, 30-hour wooden movement, dated 1840, **$375-$450.**

3. Welch, Spring & Co., Forestville, Conn., round top, full pillars, 8-day, time and strike, Pat. 1868, **$165-$190.**

4. American-made, carriage-type alarm, Pat. 1887-1901, **$85-$110.**

5. Ansonia Clock Co., double figure mantel clock, bronze figures, ''Music and Poetry,'' 8-day, time and strike, **$420-$480.**

6. Ansonia Clock Co., ''Racket'' alarm clock, strikes on hour and half hour, **$30-$40.**

1. Wm. L. Gilbert Clock Co., Winsted, Conn., "Occidental," walnut, mirror side shelf clock, time and strike, 1871-1934, **$360-$450.**

2. Wm. L. Gilbert Clock Co., Figure 8 wall clock, "Janeiro," walnut and chestnut veneer, time and strike, **$395-$465.**

3. C. & N. Jerome, Bristol, Conn., mahogany 30-hour brass weight, half round hollow columns, 1839-40, time and strike, **$420-$500.**

4. E. N. Welch Mfg. Co., Forestville, Conn., walnut shelf clock, time, strike and alarm, **$240-$285.**

5. E. N. Welch Mfg. Co., OG, 30-hour spring movement, time and strike, **$195-$290.**

1. E. N. Welch Mfg. Co., Forestville, Conn., walnut kitchen clock, time and strike, 1864-1903, **$250-$320.**

2. E. N. Welch, chestnut shelf clock, "Coghland," time, strike and alarm, 1864-1903, **$250-$310.**

3. E. N. Welch, pressed oak shelf clock, "Robert E. Lee," time and strike, 1864-1903, **$320-$410.**

4. F. Kroeber Clock Co., New York City, miniature timepiece, **$72-$92.**

5. Ansonia Clock Co., brass table clock, 8-day, Pat. 1892, **$110-$160.**

6. Ansonia Clock Co., Crystal Regulator, 8-day, time and strike, mercury pendulum, open escapement, **$250-$325.**

7. E. Ingraham & Co., Bristol, Conn., "Doric" rosewood mantel clock, 8-day, time and strike, Pat. 1871, **$225-$295.**

8. American-made, miniature cottage clock, time and alarm, **$125-$165.**

9. Seth Thomas, cottage clock, rosewood case, 8-day, strike and alarm, **$240-$290.**

10. Wm. L. Gilbert Clock Co., Winsted, Conn., miniature shelf clock, Octagon Top, rosewood case, 1866-1871, time only, **$145-$175.**

1. E. N. Welch Mfg. Co., Forestville, Conn., walnut teardrop mantel clock, time and strike, **$255-$355.**

2. Samuel Curtiss, Unionville, Conn., Flat OG, 30-hour wooden movement, time and strike, **$285-$345.**

3. Waterbury Clock Co., Waterbury, Conn., metal cased kitchen type clock, Pat. 1874-5, time and strike, **$160-$225.**

4. American-made "Keyless" rim wind, rim set, 8-day, Pat. 1907-1915, **$82-$95.**

5. Ansonia Clock Co., black cast iron, 8-day, time and strike, **$120-$165.**

6. Ansonia Clock Co., Royal Bonn porcelain case, open escapement, time and strike, **$265-$355.**

1. Gilbert Mfg. Co., Winsted, Conn., gilt and shell column, 30-hour, brass weight movement, time and strike, **$220-$275.**

2. F. Kroeber Clock Co., New York City, walnut teardrop shelf clock, 8-day, time and strike, 1850-1880s, **$260-$325.**

3. Seth Thomas, miniature Empire, full pillars, 8-day, time, strike and alarm, ST hands (''S'' on 1 hand, ''T'' on other hand), **$270-$325.**

4. New Haven Clock Co., walnut shelf clock; has ''Jerome & Co.'' label, 8-day, time and strike, **$225-$285.**

5. Seth Thomas, metal cased, 8-day, time and strike, **$275-$365.**

6. Ansonia Clock Co., Gothic, mahogany (steeple), 8-day, time, strike and alarm, **$365-$425.**

1. Ansonia Brass & Copper Co., Ansonia, Conn., rosewood case, sharp Gothic, 8-day, time and strike, 1851-1878, **$240-$285.**

2. Ansonia Clock Co., walnut shelf clock, "King," 8-day, time and strike, **$440-$485.**

3. Daniel Pratt & Co., Reading, Mass., mahogany Beehive, 8-day, time and strike, 1832-46, **$325-$395.**

4. Gilbert Mfg. Co., rosewood case, steeple, 30-hour, time and strike, dated 1868, **$185-$265.**

5. New Haven Clock Co., bronze single figure mantel clock, 8-day, time and strike, **$350-$425.**

6. Seth Thomas, Round top shelf clock with full pillars, 8-day, time, strike and alarm, **$225-$275.**

1. E. Ingraham & Co., Bristol, Conn., oak kitchen clock with calendar dial, "Gila," from their "River" line, time and strike, **$320-$425.**

2. Ithaca Clock Co., Ithaca, N. Y., No. 8, shelf library calendar, walnut, 30-day, double spring time movement, H. B. Horton's patents, time and strike, **$1,300-$1,650.**

3. Ansonia Clock Co., oak kitchen clock, time, strike and alarm, **$225-$285.**

4. F. Kroeber Clock Co., New York City, alarm clock, **$20-$28.**

5. Waterbury Clock Co., Waterbury, Conn., black cast iron, porcelain dial, 8-day, time and strike, **$135-$168.**

6. Waterbury Clock Co., miniature schoolhouse clock, 8-day, time only, **$175-$215.**

7. Wm. L. Gilbert Clock Co., Winsted, Conn., single figure mantel clock, "Hercules," cast metal, visible escapement, 8-day, time and strike, **$325-$410.**

1. Seth Thomas, Plymouth Hollow, Conn., mahogany OG, 30-hour, brass weight movement, time and strike, **$240-$320.**

2. Seth Thomas, Plymouth Hollow, Conn., rosewood, 30-hour, brass weight movement, "Buckingham Palace" on tablet in door, time and strike, **$225-$285.**

3. H. Welton & Co., Terryville, Conn. (successors to Eli Terry & Co.), mahogany OG, 30-hour, brass weight movement, time and strike, **$230-$285.**

4. Atkins Clock Co., Bristol, Conn., round top shelf clock with pillars, 8-day, time and strike, Pat. 1867, **$220-$275.**

5. Ansonia Clock Co., alarm clock with 24-hour setting, **$51-$63.**

6. Ansonia Clock Co., single figure mantel clock, bronze figure, "Attila," 8-day, time and strike, **$700-$800.**

7. Gilbert Clock Co., "Anniversary," black wood mantel clock, brass bell on top of clock, "strikes" time, 8-day, **$210-$285.**

1. E. N. Welch Mfg. Co., Forestville, Conn., oak kitchen clock, time, strike and alarm, **$210-$260.**

2. Ithaca Clock Co., rosewood case, No. 4 hanging office clock, 30-day, double spring, nickel-plated movement, 1866, H. B. Horton's patents, **$1,350-$1,550.**

3. Wm. L. Gilbert Clock Co., Winsted, Conn., walnut Victorian shelf clock, 8-day, time and strike, **$425-$485.**

4. German, Round top mantel clock, Westminster chime, **$140-$170.**

5. E. Ingraham & Co., Bristol, Conn., "Nomad" mantel clock, 8-day, time and strike, **$72-$85.**

1. Seth Thomas, walnut, Drop Octagon schoolhouse clock, 8-day, time and strike, **$325-$400.**

2. Seth Thomas, walnut kitchen-type wall clock, 8-day, time, strike and alarm, **$235-$275.**

3. Waterbury Clock Co., rosewood, Drop Octagon, 8-day, time and strike, **$290-$350.**

4. Brewster & Ingraham, Bristol, Conn., steeple, 30-hour, time and strike, brass springs, 1844-1852, **$225-$275.**

5. E. N. Welch, Forestville, Conn., Iron Front, 8-day, time and strike, **$185-$225.**

1. Atkins Clock Co., Bristol, Conn., rosewood round top wall clock, time and strike, 1859-79, **$300-$420.**

2. Seth Thomas, Parlor Calendar No. 1, rosewood, weight driven, Pats. 1854-57, 1860-62, time and strike, **$1,300-$1,650.**

3. Daniel Pratt, Son, Reading, Mass., rosewood round top wall clock, time only, 1860-75, **$280-$320.**

4. Ansonia Clock Co., oak kitchen-type hanging clock, mercury pendulum, time and strike, **$245-$265.**

5. New Haven Clock Co., oak, mirror side shelf clock, time and strike, **$270-$345.**

1. Wm. L. Gilbert Clock Co., Winsted, Conn., walnut wall clock, "Stardrop," 8-day, time and strike, **$540-$635.**

2. Waterbury Clock Co., Waterbury, Conn., walnut, spring movement, wall clock, 8-day, time and strike, **$555-$635.**

3. Seth Thomas, oak, World Regulator, 15-day, double spring, time and strike, **$525-$725.**

1. Terry Clock Co., Pittsfield, Mass., Octagon Drop wall calendar clock, 8-day, 1880s, time only, **$385-$475.**

2. Seth Thomas, walnut clock, single weight, time only, **$420-$510.**

3. Ansonia Clock Co., oak Drop Octagon, time only, **$250-$320.**

1. E. Ingraham & Co., Bristol, Conn., Round Drop wall clock, calendar, 8-day, Pat. 1872-73, time only, **$410-$495.**

2. E. Ingraham & Co., Round Drop wall clock, time only; date on lower glass: June 1, 1887, **$500-$610.**

3. Ansonia Clock Co., single figure bronze mantel clock, "Shakespeare," 8-day, time and strike, **$370-$400.**

4. Ansonia Clock Co., single figure bronze mantel clock, "Philosopher," 8-day, time and strike, **$380-$475.**

1. Welch, Spring & Co., Bristol, Conn., B. B. Lewis Regulator Calendar No. 2, rosewood, mechanical, double weight, 1864, 8-day, time and strike, **$1,600-$1,975.**

2. American-made, mahogany case, stenciled pillar and splat, 30-hour, wood, groaner-type movement, bell on top, time and strike, **$245-$290.**

118

Additional Clocks with Prices

(Not Illustrated)

Page numbers of all clocks listed in this CLOCK GUIDE
may be found by referring to the Index on page 147.

AMERICAN-MADE Many wholesale houses, jewelers, mail order houses, sold clocks that had a trade name but no maker's name. Here are just a few of the thousands sold throughout America from the 1850s until World War II. Obviously, we're wrong on some of these and for that we apologize.

Iron & Bronze:

Sambo, 1-day, winker, iron ("winker"—the eyes move)	**$960-1,150.**
Bouquet, 1-d., iron	**$145-175.**
Bouquet, 8-day, iron, Fancy	**$200-250.**
Horse, 8-d., bronze	**$220-265.**
Topsey, 1-d., winker, iron	**$960-1,125.**
Globe, 8-d., S., bronze	**$210-250.**
Continental, 1-d., winker, iron	**$1,100-1,400.**
Reaper, 8-d., s., bronze	**$200-240.**
Eagle, 8-d., S., bronze	**$200-240.**
Etruscan, 8-d., bronze	**$215-235.**
Lion Head, 1-d., S., bronze	**$200-250.**
Oak Leaf, 8-d., S., bronze	**$210-265.**
Drama, 1-d., S., bronze	**$155-175.**
Same except 8-d.	**$160-180.**
Washington, 1-d., S., iron	**$155-180.**
Same except 8-d.	**$210-235.**
Lattice, 1-d., S., iron	**$220-255.**
Same except 8-d.	**$230-270.**
Mustang, 8-d., S., bronze	**$220-260.**
Parlor, 1-d., S., iron	**$135-$160.**
Same except 8-d.	**$155-$178.**
Opera; Patchen, 8-d., S., bronze	each **$220-265.**
Temple, 1-d., S., iron	**$210-270.**
Same except 8-d.	**$225-265.**
Arbor, 1-d., S., iron	**$165-195.**
Same except 8-d.	**$225-270.**
Webster, 8-d., S., iron	**$145-185.**
Birds, 8-d., S., bronze	**$220-260.**
Wine Drinker; Gleaner; Drummer; Vintner; Dragon; Guardian; Juno, 8-d., S., bronze	each **$175-222.**
Doric, 1-d., iron	**$125-145.**
Dolphin, 8-d., S., bronze	**$215-240.**

Alarms:

Parlor, walnut case, gold trimming	**$52-65.**

Fire Alarm, copper case, gold plated dial	$67-80.
Jumbo Watch Alarm Clock, nickel case	$62-80.
Sleigh-Bell, copper case, gold gilt	$53-82.
Patrol Alarm, 1-d., luminous dial	$40-58.
Same except plain dial	$44-54.
New Echo, nickel plated	$33-48.
Tiny Time, nickel plated, 1-d., T.	$27-34.
Boom; Queen; Bell, 1-day	each $32-50.
Pert, 1-d., Lever	$56-72.
Same except T. only	$47-61.
Black & Tan, 1-d.	$56-72.
Mandolin, 1-d.	$57-66.

Novelties:

Climax, nickel, 1-d., T., alarm	$120-162.
Carriage, extra nickel, 1-d., ½-Hr. S. & Alarm	$235-278.
Lurine, 1-d., T & alarm, nickel	$156-178.
Duke, nickel or gilt, 1-d., T	$42-61.
Fairies, Rich Gold, Antique Brass, 1-d., T.	$120-160.
Little Corrine, nickel sides, glass top, 1-d., T.	$80-110.
Jockey, silver or bronze finish, 1-d. T.	$98-152.
Castle, silver finish, 1-d., T	$80-97.
Basket, silver finish	$63-78.
Barge, silver finish	$140-162.
Lyre, silver finish, 1-d., T	$52-70.
Conductor, gold, silver or oxide silver finish, 1-d., T & A	$120-162.
Same, plus ½-Hour Strike	$140-165.
Empire, Rich Gold or Ant. Brass finish, 1-d., T	$110-143.
Chef, fancy metal finish, 1-d., T	$62-81.

ANSONIA CLOCK CO., Ansonia, Connecticut, founded by Anson V. Phelps, 1851-1878; then Brooklyn, New York, 1879- c. 1930s. Clocks listed here are 1880s to 1920s.

Automobile clocks:

Motor, Arabic or Roman dial, brass or nickel, 1-day, stem wind	$45-60.
Automobile, Arabic or Roman dial, black & nickel, 8-day	$46-56.
Automobile #1, Arabic or Roman dial, black & nickel, 8-day	$47-63.
Automobile #2, black, 8-day	$38-58.

Nickel alarms:

Bee, Arabic or Roman dial, 1-day, winds by turning the back	$61-71.
Bee, Arabic or Roman dial, 8-day, winds by turning the back	$69-82.
Spark, Arabic or Roman dial, 1-day	$37-51.
Spark, Arabic or Roman dial, 8-day	$72-82.
Princess, Roman dial, 1-day	$44-56.
Princess, Roman dial, 8-day	$64-72.
Pirate, Roman dial, 1-day	$32-50.
Luminous Pirate, Roman dial, 1-day	$51-63.
Clatter, Arabic or Roman dial, 1-day	$28-37.

Racket, Roman dial, 1-day	$60-70.
Amazon, Arabic or Roman dial, 1-day	$62-73.
Rouser, Arabic or Roman dial, 1-day	$32-43.

Japanese Bronze alarms:

Bronto, Arabic or Roman dial, 1-day	$82-91.
Rattler, Arabic or Roman dial, 1-day	$79-86.

Swinging Ball clocks:

Arcadia; Diana; Fortuna; Gloria; Hunter; Juno. All are Arabic dial, real bronze finish, 8-day	each $1,650-$1,840.
Double Figure, Arabic dial, real Bronze Finish, 8-day	$2,500-2,750.

Single Figure Bronze clocks:

Columbia, Syrian Bronze Finish, Cathedral gong, ½-Hour Strike, 8-day, Porcelain Visible Escapement dial	$1,550-1,850.
Mars, Syrian or Japanese Bronze Finish, etc.	$750-900.
Pizarro, Syrian or Japanese Bronze Finish, etc.	$625-755.
Tasso, Syrian or Japanese Bronze Finish, etc.	$585-676.
Lothario, Syrian or Japanese Bronze Finish, etc.	$585-690.
Don Juan, Syrian or Japanese Bronze Finish, etc.	$690-925.
Insult, Japanese Bronze, etc.	$580-675.
Fisher, Japanese, Barbedienne or Syrian Bronze Finish, etc.	$710-825.
Troubadour, Japanese Bronze Finish, etc.	$485-525.
Opera, Japanese or Syrian Bronze Finish, etc.	$485-625.
Industry, Japanese Bronze Finish, etc.	$440-620.
Mozart, Gilt, Japanese, Barbedienne or Syrian Bronze Finish, etc.	$495-570.
Reubens, Japanese Bronze Finish, etc.	$482-525.
Cincinnatus, Japanese Bronze Finish, etc.	$480-610.
Fantasy, Japanese Bronze Finish, etc.	$420-585.
Arion, Composer, Macbeth, Knight, Real Bronze Finish, etc.	each $495-575.
Siren, Hermes, Rex, Japanese Bronze Finish, etc.	each $465-587.
Don Caesar, Japanese Bronze Finish, etc.	$595-695.

Double Figure Bronze clocks:

Vocalists, Syrian Bronze Finish, etc.	$1,250-1,675.
Art & Commerce, Pizarro & Cortez, Combatants, Muses, Don Caesar & Don Juan, Japanese Bronze Finish, etc.	each $1,250-1,475.
Fisher & Hunter, Japanese or French Bronze Finish, etc.	1,350-1,475.

Crystal Regulators:

Oriel, Polished Brass or Gold Plated, visible escapement, 8-day	$400-475.
Coral, same as above	$250-375.
Claudius, same as above	$265-295.
Rouen, same as above	$275-299.
Wanda, same as above	$320-366.
Vulcan, Rich Gold, etc.	$280-310.
Cetus, Clifton, Polished Brass or Gold Plated, etc.	$300-362.
Elysian, Rich Gold, etc.	$675-922.
Eulogy, Rich Gold, etc.	$380-450.
Jupiter, Polished Brass, Rich Gold Ornaments, Hour & ½-Hour Gong Strike, 8-day	$1,600-$1,950.

Porcelain Regulators:

Porcelain Regulator No. 2, Royal Bonn—decorated Top & Base, etc.	**$1,475-$1,785.**
No. 4 & No. 5, same as above	each **$1,600-1,900.**

Regulators:

Capitol No. 102, Spring Time, Roman dial, finished in black walnut or ash	**$1,100-1,350.**
Capitol No. 103, Gong Strike, etc.	**$1,250-1,425.**
Capitol No. 104, Weight, Time, etc.	**$1,400-1,575.**
Santa Fe No. 100, Roman dial, Weight, Time, black walnut, mahogany or oak	**$1,375-1,565.**
Santa Fe No. 101, Roman dial, ½-hour, Gong Strike, mahogany, or mahogany and ash	**$1,300-1,450.**
Standard, Roman or Arabic dial, Weight, Time, 8-day, mahogany, or oak, polished	**$600-750.**
Bagdad No. 108, Roman or Arabic dial, Spring, Time, black walnut, mahogany, oak or ash	**$785-995.**
Bagdad No. 109, same except Gong Strike	**$825-895.**
Bagdad No. 110, same except Weight, Time	**$1,400-1,675.**

Office Regulators:

Pacific No. 27, oak, 8-day, Time	**$785-$925.**
Pacific No. 28, same except Strike	**$700-795.**
Pacific No. 29, same except Time, Calendar	**$875-950.**
Pacific No. 30, same except Strike, Calendar	**$900-1,100.**

Black Enameled Iron Mantel clocks:

"B" Assortment, 8-day, Hour & ½-hour Gong Strike, Arabic or Roman dial, Japanese Bronze Finish, choice of American Sash, French Sash, visible escapement, etc. — Bangor; Batavia; Bath; Butte; Brandon	each **$120-$155.**
"C" Assortment, same as "B" — Cambridge; Carlisle; Chatham; Compton; Coventry	each **$110-150.**
"Paris" Assortment, same as "B" & "C" — Chester; Paris; Venice; Palermo; Vienna; Sorrento	each **$100-138.**
"Montague" Assortment, same as above, except Gilt or Barbedienne Bronze Ornaments — Belgium; Hague; Rosalind; Sevres	each **$92-127.**

Novelty clocks:

Breton; Good Morning (#268); Good Night (#254); Normandie; Arabic or Roman dial, 8-day, Porcelain dial, Bronze Finish	each **$92-120.**
Amigos; Elf; Eveline; La Belle; La Rose; La Reine, same except Rich Gold Finish	each **$78-99.**

Calendars:

Octagon Long Drop, wall, oak, 8-day, Simple Calendar	**$495-610.**
Octagon Long Drop, wall, walnut, 8-day, Simple Calendar	**$625-735.**
Victorian Kitchen, Time & Strike, 8-day, oak, barometer, thermometer	**$320-380.**
Desk Inkwell, 1-day, Simple Calendar	**$320-420.**

WM. L. GILBERT CLOCK CO., Winsted (Winchester), Connecticut: c. 1850-1866, Wm. L. Gilbert Co.; 1866-1871, Gilbert Mfg. Co.; 1871-1934, Wm. L. Gilbert Clock Co.; 1934-1957, Wm. L. Gilbert Clock Corp.; c. 1964, business sold to Spartus Co., Chicago.

Shelf clocks:

Eros; Prince; Hestia; Dio; Laurell; Mahuta; Edina; Peto; Cipango; 8-day, Strike; 8-day, Strike, Cathedral Gong, walnut	each **$170-210.**
Abyla; Werra; Calpe, same except ash with walnut trimming	**$185-235.**
Crius, same except ash	**$175-230.**
Flora, same walnut with ash trimming	**$170-245.**

Crystal Regulators:

Tuscan, 8-day, ½-Hour Strike, Ivory Porcelain Dial, visible escapement, mercury pendulum	**$825-910.**
Vista, same as above	**$275-340.**
Venice, same as above	**$725-825.**
Trinity, same as above	**$695-765.**
Verdi, same as above	**$995-1,250.**
Terese, same as above	**$785-890.**
Tunis, same as above	**$620-695.**

Alarms:

365 Nite; Tornado; Bi-Nite; San Toy; black dial, luminous, Nickel Finish, 40-Hour	each **$42-56.**
La Sallita; Tiger; Nickel Plated Brass case, 1-day	each **$44-60.**
Iron Clad, 1-day, Heavy Bronze Plated Case	**$41-58.**

Regulators:

No. 9 Hanging, walnut, cherry, ash or oak, glass sides, 8-day, Weight, Time	**$2,600-2,900.**
No. 10, Hanging, same as above	**$875-965.**
No. 9 Standing, same as above	**$2,600-2,900.**
"A", walnut only, same as above	**$495-675.**
Bonita, 8-Hour, Spring Pendulum, Mahogany Flat Finish	**$325-395.**
Leeds, same as above	**$275-350.**

Mantel clocks:

Refined; Big Premium; Hand-Rubbed Mahogany, 40-Hour, Time	**$67-82.**
Junior; Two-Color; Orlando; Cornell; Sheridan; Exquisite; Hatherly; Newark; Mobile; Debutante; Beckworth; Ludlow; Kimberley; Pompton; Astoria; Bloomfield; Osborne; Roscoe; Hand-Rubbed Mahogany, 8-day, Pendulum, Strike	**$82-102.**

Calendars:

Victorian Kitchen, walnut, 8-day, Time & Strike, Spring	**$720-780.**
Eclipse Regulator, oak, 8-day, Spring, Simple Calendar	**$570-580.**
Octagon Drop, wall, oak, 8-day, Spring, Simple Calendar	**$420-455.**
Eureka, oak, 8-day, Time & Strike, Spring, Simple Calendar	**$467-495.**
National, oak, 8-day, T & S, Spring, Simple Calendar, thermometer and barometer	**$695-740.**

Oriental, oak or walnut, 8-day, Gong   **$1,350-1,625.**

E. INGRAHAM & CO., Bristol, Connecticut. Elias Ingraham founded the firm in 1805. 1880-1884, E. Ingraham & Co.; 1884-1958, The E. Ingraham & Co.; 1958, The E. Ingraham Co.; present, The Ingraham Co.

Banjos:
Nordic; Norse; Neptune; Norway; Nile; #1; #2; Circular
    Finish, Silver Plated Dial, 8-day, Marine Movement,
    mahogany, colors   each **$125-175.**
Hyannis; Wellfleet, same as above   **$250-310.**
Treasure, same except Hand Rubbed Mahogany &
    Pendulum Movement   **$485-525.**

Cabinet clocks:
Press; Globe; Sun; Ingot; Puck; World, 1-day & 8-day   each **$162-193.**
Bullion; Ingot; Times; World; Rondo, 8-day, ½-Hour
    Strike   each **$190-265.**
Acme; Era; Nos. 7, 8, 9, 10, 8-day, ½-Hour Strike, Gong   each **$148-179.**
Tulip; Dahlia; Violet; Lilac, oak or walnut, 8-day, Strike   **$163-172.**

Mantel clocks:
Berlin; Belmont; Burgundy; Hermes; Hampton;
    Hammond; Hather; Magnet; Howard; Nomad;
    Magic, Mahogany Finish, 8-day, Hour Cathedral
    Gong, ½-Hour Cup Bell   each **$72-90.**

Hanging Office clocks:
Drop Octagon, solid oak, T & S, 8-day   **$220-285.**
Dew Drop, oak, rosewood or walnut, 8-day   **$260-322.**
Misay, rosewood, 8-day   **$265-285.**

Alarms:
Seville; Sancho; Sibl; Saturn, 8-day, Solid Mahogany
    Case   each **$68-83.**
Indian; Sentry; Ideal; Autocrat, Nickel, Brass Polished,
    Large Bell, 1-day (also available in 8-day)   each **$51-63.**

Calendars:
Drop Octagon, wall, solid oak, Time and/or Strike   **$328-363.**
Press, shelf, walnut, 8-day, Strike   **$175-240.**
Press, shelf, walnut, 8-day, Strike, Gong   **$220-260.**
Globe, shelf, oak, 8-day, Strike   **$195-248.**
Globe Extra, shelf, oak or walnut, 8-day, Strike,
    Thermometer and Barometer   **$187-198.**
Globe Extra, shelf, oak or walnut, 8-day, Strike,
    Alarm, Thermometer and Barometer   **$250-290.**
Parlor, shelf, walnut, Spring, T & S   **$1,450-1,725.**
Figure Eight, wall, ash, 8-day, T & S, B. B. Lewis
    Movement   **$1,550-1,820.**
Victorian Kitchen, wall, oak, 8-day, T & S, Spring,
    Simple Calendar   **$290-320.**
Gila, shelf, oak, 8-day, T & S, Spring, Simple Calendar,
    Thermometer and Barometer   **$296-380.**

Dew Drop, wall, 8-day, T & S, Spring, Simple Calendar   **$410-485.**

ITHACA CALENDAR CLOCK CO., Ithaca, New York, founded 1865 by H. B. Horton. Out of business around World War I.

No. 2 Bank, wall, walnut or ash, 8-day, Weight, Time	**$1,400-1,650.**
No. 0 Bank, wall, walnut, 8-day, Weight, Time	**$3,900-4,650.**
No. 5½ Belgrade, wall, walnut or ash, 8-day, ½-Hour Slow Strike	**$2,200-2,550.**
No. 5½ Belgrade, wall, walnut or ash, 30-day, Double Spring, Strike	**$1,975-2,600.**
No. 14 Granger, walnut, 8-day, ½-Hour, Slow Strike, Alarm	**$1,100-1,450.**
No. 2½ Brisbane, walnut or ash, 8-day, ½-Hour Slow Strike, Gong	**$2,220-2,650.**
No. 2½ Brisbane, walnut or ash, 30-day, Double Spring, Time	**$2,200-2,700.**
No. 4 Office, wall, walnut or rosewood, 8-day, ½-Hour Slow Strike	**$1,400-1,585.**
No. 4 Office, wall, walnut or rosewood, 30-day, Double Spring, Time	**$1,400-1,675.**
No. 7 Cottage, wall, walnut, 8-day, Strike	**$1,150-1,400.**
Ionic, walnut, 8-day, T & S	**$575-625.**
Alexis No. 1, walnut, 30-day, Time	**$600-710.**
Alexis No. 2, walnut, 8-day, Time	**$525-625.**
Alexis No. 3, walnut, 8-day, Strike	**$515-610.**
Library, wall, walnut, 8-day, Double Spring, Pendulum Movement	**$1,375-1,620.**
Library, wall, walnut, 8-day, Pendulum, Strike	**$1,400-1,650.**
No. 5 Round Top, walnut, 8-day, Double Spring	**$1,200-1,400.**
No. 5 Round Top, walnut, 8-day, Pendulum, Strike	**$900-1,450.**
No. 1 Regulator, wall, walnut, Sweep Second Hand, 8-day, Weight, Time	**$6,300-6,775.**
Iron Case, wall, 8-day, Spring, Weight, walnut	**$2,100-2,350.**
Large Iron Case, wall, walnut, 30-day, Double Spring, Hubbell Movement	**$1,800-2,450.**
No. 0 Regulator, walnut, 8-day, Double Weight (also Shelf Model)	**$2,500-2,750.**
No. 15 Melrose, cherry, 8-day, Strike	**$2,450-2,625.**
No. 12 Kildare, wall, mahogany, 8-day, Spring, Strike	**$2,500-2,750.**
No. 12 Kildare, mahogany, 30-day, Spring, Time, wall	**$3,475-3,670.**
No. 13 Kildare, shelf, mahogany, 8-day, Spring, Strike	**$2,200-2,540.**
No. 13 Kildare, shelf, mahogany, 30-day, Spring, Time	**$3,400-3,650.**
No. 5 Emerald, shelf, walnut, 8-day, Spring, T & S	**$1,650-1,900.**

MISSION clocks. This is not a make but a type of clock popular in the early 1900s and until c. World War I. Make of oak — solid, weathered oak, mahogany-finished oak, etc., some had brass hands, copper dials. Most were 8-day, ½-Hour Strike with Cathedral Gong.

Mantel, weathered oak, wood dial, brass hands   **$92-120.**

Mantel, solid oak, wood dial, Arabic numerals, pendulum movement	**$72-90.**
Mantel, Los Barrios; San Martin, mantel, same as above, etc.	each **$70-88.**
La Plata, same as above except stained glass in door	**$92-120.**
Denmark, standing, solid oak, 8-day, etc.	**$220-246.**
Antwerp, standing, same as above, etc.	**$140-170.**
Rondo, wall, 8-day, Arabic numerals, Lever, etc.	**$50-70.**
Madrid, wall, brass hands & figures, 8-day, ½-Hour, etc.	**$70-92.**
Dining Room, wall, light oak, copper dial, 1-day, double weight, etc.	**$172-192.**
Den, wall, dark oak, 8-day, T.	**$98-127.**

NEW HAVEN CLOCK & WATCH CO., New Haven, Connecticut. Founded in 1853 by Hiram Camp. Originally made movements for the Jerome Mfg. Co. Firm enlarged; electrics made in 1920; firm went out of business in 1965.

Enameled Irons:

Plain White Gilt or Pearl dials, 8-day, ½-Hour Strike, Cathedral Gong, Black, Smoke or Malachite Finish — Rosalind; Fayette; Mona; Chiselhurst; Leland; Scandia; Chateau; Galatea; Thetis; Ferncliff; Carnival; Titan; Fairfield; Colonna; Washington	each **$150-167.**

Enameled Blackwoods:

Same as Irons. A Music Box attachment was available: Malvern; Seneca; Pembroke; Alexander; Clarendon; Bingen; Loring; Prescott; Numa; Mantua; Dante; Bancroft; Lohengrin; Burlington	each **$88-115.**

Decorated Porcelains:

8-day, Hour & ½-Hour Strike, Cathedral Gong — Herbert; Hamilton; Mabel; Rosendale; Malabar	each **$195-222.**
Horican; Holly; Haverford; Gerald, same as above, have V.E.	each **$400-465.**
Fleetwood; Mortan; Wakefield; Lionel, 1-day, Time, usual markings	each **$52-72.**

Alarms:

Nickel, 1-day — Mauser; Beacon; The Fly; Sprite	each **$32-42.**
Tatoo, Nickel Plated, Seamless Brass case, 1-day	**$32-41.**
Jr. Tatoo, 1-day, Satin Silver, Rich Gold, Gun Metal Finishes	**$50-59.**
The Commuter; Brownie; Tocsin; Beacon Luminous; Beacon	each **$40-51.**
Sting, Nickel, 1-day, Assorted Colors	**$37-42.**
Herald, wood case, Nickel or Brass front, 1-day	**$36-50.**
Sultana, Nickel, 1-day, Alarm; 1-day, Strike	**$29-39.**
Elfin, Nickel, 1-day, T., A.; 1-day, S., Fancy Dial	**$35-39.**
Alert, Nickel, 1-day, A.; 1-day, T.A., Plush	**$40-60.**

Single Figure (Bronze or Silver Finish):

8-day Gong Strike, Visible Escapement, Finish does not effect value of clock.

Marine; Luli; Philip	each **$525-675.**
Sir Christopher; Rebecca; Infantry	each **$480-562.**
Clovis	**$520-585.**
Mignon; Brennus	each **$480-562.**
Fisher Boy	**$650-725.**
Don Juan; Don Caesar; Hunter & Dog; Carthage	each **$562-615.**

Single Figure (Gilt or Bronze Finish):
Same as above; Finish does not effect value of clock.

Fame; Octavius; Bernard Palissy; Poetry; Ivanho; Horse; Flower Girl; Clotho; Norman; Benvento Cellini; Flute Player; Knight; Saxon; Roman	each **$480-522.**

Mantels:

Picket Line; Patrol Line; Maine Line; "D" Line, oak, 8-day, ½-Hour Strike; ½-Hour Strike, Alarm; ½-Hour Strike, Gong	each **$150-162.**
Same clocks in walnut	each **$170-230.**
Metal Trimmed, Brazilian Line; Merchant's Line; Felix; Forum, same as above. In oak	each **$146-156.**
In walnut	each **$160-170.**

Veneers:

O.G. (Weight #2), 1-day, Strike; 1-day, Strike, Alarm	**$220-245.**
O.O.G. (Weight), same as above	**$218-257.**
Sharp Gothic "A", 1-day, S.; 1-day, S.&A.	**$118-135.**
Sharp Gothic "A", 8-day, S.; 8-day, S.&A.	**$160-186.**
Sharp Gothic "G", same as above	**$119-153.**
Duchess, 8-day, Spring; 8-day, Spring, Alarm	**$150-170.**
Round Gothic, 1-day, Strike, Spring	**$140-160.**
Round Gothic, 8-day, Strike	**$172-182.**
Round Gothic, V.P., 1-day, Strike, Spring	**$140-172.**
Round Gothic, V.P., 8-day, Strike, Spring	**$158-172.**
Tuscan, V.P., 1-day or 8-day, Strike, Spring	**$100-120.**
Cottage, 1-day, Time, Spring; 1-day, Strike	**$87-100.**
Cottage, Extra, 1-day, etc.	**$100-110.**
Cottage, Extra, 8-day, etc.	**$115-130.**
Guide, 1-day, etc.	**$120-140.**
Guide, 8-day, etc.	**$140-160.**
Gothic, Gem, V.P., 1-day, etc.	**$120-150.**
Gothic, Gem, V.P., 8-day, etc.	**$142-173.**

Office:

Regulator, wall, walnut veneer, solid walnut circle, 8-day, T.; 8-day, T & S	**$325-345.**
Regulator, Blake, wall, solid oak, 8-day, T & S	**$220-260.**
Regulator, Braddock, wall, solid oak, 8-day, T & S	**$195-220.**
Bank, oak, 8-day, T & S	**$270-299.**
Vamoose, wall, solid oak, 8-day, T & S, Gong	**$525-595.**
Grecian, wall, solid oak, 8-day, T & S, Gong	**$650-725.**

Jeweler's Regulators:

Thornton, wall, oak or cherry, 8-day, Swiss movement, Sweep Second	**$2,200-2,900.**

New York, wall, solid oak, same as above $2,600-2,795.
Giant, standing, solid oak, same as above $3,450-3,775.
Chippendale, standing, oak or mahogany, same as above $2,800-3,400.

Mantels, Black Walnut

(1-day, Strike; 1 day, Strike, Alarm):

Christine; Bobolink; Clochette; Bonita; Lirus; Charmer;
    Calumet; Ivy; Nereid; Boreas; Anita   each $200-245.
Irex; Rambler; Cecile; Celest; Elna; Mayflower;
    Cinderella; Vindex; Osage; Coquette; Cygnet;
    Volante; Zingara; Vampire; Parisian —
        8-d., S   each $220-265.
        8-d., S & Alarm   each $240-280.
        8-d., Cathedral Gong S.   each $240-270.
        8-d., Cath. Gong S. & Alarm   each $240-282.

SESSIONS CLOCK CO., Forestville, Connecticut. Sessions took over the E. N. Welch Mfg. Company operations in 1903.

Banjos:

Halifax; Provincetown; Salem, mahogany case, 8-d.
    Lever movement, Eagle & Brackets, etc., c. 1932   each $295-320.
Revere, mahogany finish case, T. $280-322.
Same except T & S $285-340.
Hyannis, Old Ivory Finish case, 8-d., Lever $250-265.
Wellfleet, Green Lacquer Finish, 8-d., Lever $250-295.
Lexington; Narraganset   each $216-272.
Star Pointer Regulator, quartered oak, 8-d., T $275-315.
Same as above except with Calendar $370-410.
Gallery, Wall, 8-d. $160-175.
Regulator No. 2 $279-400.
Regulator No. 3 $475-550.

Enameled Blackwoods:

Arcadia; Goldenrod; Goldstar; Marbleized Nos. 1 & 2;
    Marigold; Baldwin; Ardmore; Mozart; Manhattan;
    Elton; Corinth; Ideal; Melrose   each $160-185.

SETH THOMAS CLOCK CO., (Plymouth Hollow), Thomaston, Connecticut. Founded by Seth Thomas in Plymouth Hollow in 1853. Made Southern Calendar movements in 1875. Town name changed to Thomaston in 1866. Made watches from 1883 to 1914. Firm became part of General Time Instruments Co. — The General Time Corp. — after 1949. Firm still in business.

Regulators: (All are 8-day, Weight, Time and/or Strike)

No. 2, walnut, cherry or old oak veneer, polished $775-$890.
No. 3, walnut, rosewood, old oak veneer, polished,
    Strike $1,400-1,575.
No. 6, walnut, cherry, oak or old oak veneer, polished $1,600-1,800.
No. 8, walnut $1,750-1,900.
No. 9, walnut, oak, old oak veneer, polished $1,500-1,750.

Fine, No. 14, walnut case, mercury pendulum, Sweep Second	**$3,450-3,750.**
Fine, No. 15, walnut case, mercury pendulum, Independent Second	**$2,900-3,450.**
No. 16, walnut, oak or old oak veneer, polished	**$2,700-3,100.**
No. 17, walnut, cherry, old oak veneer, polished	**$1,400-1,600.**
No. 18, walnut, cherry, old oak veneer, polished	**$1,800-1,975.**
No. 19, walnut, oak or old oak veneer, polished	**$3,200-3,600.**
No. 30, walnut, cherry, oak, old oak veneer, polished	**$1,200-1,450.**
No. 31, same as above	**$1,700-1,950.**

Office:

10 ″ Drop Octagon, rosewood, walnut veneer, Gilt Lines, 8-d., Sprg., T.; 8-d., Sprg., S.	**$240-275.**
10 ″ Drop Octagon, old oak, without Gilt Lines, same works as above	**$255-276.**
12 ″ Drop Octagon, same as above	**$220-245.**
No. 3, rosewood or walnut veneer, polished, 8-day, Spring, S.	**$295-365.**
Globe, rosewood, walnut, old oak, etc., 8-day, Spring, T.	**$320-360.**
Globe, same as above except Wire Bell	**$300-325.**
Queen Anne, walnut, cherry, oak, old oak, etc., 8-d., Sprg., S., no Second Hand	**$525-675.**
Queen Anne, same as above Cathedral Bell	**$560-720.**
Flora, walnut, oak, ebony, mahogany, cherry, case hand carved, 8-day, Weight, Spring, Cathedral Bell	**$1,200-1,450.**
Jupitor, mahogany, old oak veneer, polished, dial shows Moon Phases, 8-d., S., Cathedral Bell	**$1,950-2,400.**
Lunar, same as above	**$2,600-2,875.**
No. 1, walnut, oak, 8-day, Spring, T & S	**$320-365.**
Greek, walnut, ebony, oak, mahogany, 8-day, Spring, S.	**$275-328.**
Signet, walnut, 8-day, Spring, T.	**$260-295.**
Signet, walnut, 8-day, Spring, S.	**$310-355.**

Alarms:

Echo, Elk, Nickel Plated, 1-day, T.	**$48-58.**
Tocsin, Metal case, T.	**$49-63.**
Nutmeg, Nickel Plated, 1-day	**$69-78.**

Calendars:

Parlor No. 6, walnut, oak, old oak, 8-day, Spring, Strike, Cup Bell, Perpetual Calendar	**$1,275-2,000.**
Office No. 8, walnut, oak, 8-day, Weight, Time	**$3,300-3,850.**
Parlor No. 7, walnut, hand carved case, 8-day, Spring, Strike	**$1,200-1,485.**
Office No. 10, walnut, cherry, oak, old oak, 8-d., T., Perp. Cal.	**$1,900-2,450.**
Office No. 11, same as above except mahogany instead of cherry	**$1,600-1,900.**
Office No. 12, same as No. 10	**$1,905-2,645.**

Parlor No. 10, walnut, 8-day, Weight, Time, Perpetual
    Calendar   **$1,700-2,250.**
Parlor No. 3, walnut veneer, polished, 8-day, Spring,
    Strike   **$925-$1,075.**

Levers:

Student, Mikado, Lodge, Artist, Joker, Crystal; Nickel or Gold Gilt cases or fronts	each **$140-165.**
Banner, metal case, brass or nickel plated, 8-day, T & S	**$185-220.**
Engine, 1-day, T., same cases as above	**$160-178.**
Engine, 8-day, T., same as above	**$180-198.**
Wood, mahogany or old oak veneer, polished, 8-day, T & S	**$220-260.**
Wood, same as above except 1-day, T & S	**$180-195.**
Navy, metal case, etc., 1-day, T.	**$266-287.**
Switch-Board, metal case, etc., 8-day, Time, Double Spring	**$328-369.**
Harbor, same case as above, 1- & 8-day, Strike	**$270-298.**
Chronometer, 1-day	**$290-320.**
Chronometer, 8-day	**$360-390.**

Eight-Bells:

Ship's, metal case, brass or nickel plated, 1-day, Strike	**$360-390.**
Cabin, same as above	**$270-280.**
Yacht, same as above	**$267-320.**
Boat, same as above	**$390-465.**
Wardroom, same as above	**$340-395.**

Shelf, Black Walnut:

Summit; Ogden; Alton; Omaha; Athens; Cairo; Peoria; Topeka; Reno; Albany, 8-day, Spring, Strike, Alarm, Cathedral Bell	each **$195-240.**
Princeton; Concord; Utica; Newark, same as above	each **$225-270.**

Shelf, Oak or Walnut:

New York; Harvard; Cambridge; Yale; Oxford; Cornell, 8-d., ½-H., S.	each **$150-178.**

Iron & Bronze:

Egyptian No. 1, Verde bronze, 8-day, Strike	**$350-375.**
Egyptian No. 2, Verde, Light Verde, French Bronze, 15-day, 1 & ½-Hour Strike	**$320-365.**
Egyptian No. 3., same as above	**$320-380.**
Esmerald, Bronze, 8-day, Strike	**$225-285.**
Dexter, same as above	**$265-320.**
Leisure, same as Egyptian No. 2 & 3	**$420-485.**
Huntress, same as above	**$565-680.**

Metal Cases (Gold Plated & Lacquered, Beveled Plate Glass, front & sides):

Empire No. 0, polished gold finish, 8-d., ½-H. S., C. Bell	**$320-365.**
Empire No. 2, same as above	**$450-625.**
Empire No. 4, Bronze top & base, 8-d., ½-H. S., Cath. Bell	**$350-410.**
Empire No. 5, same as above	**$550-700.**
Empire No. 9, Rich Gold Finish, 8-day, etc.	**$495-625.**
Empire No. 15, same as above	**$525-625.**

Empire No. 19, same as above & Burnished	**$625-650.**
Empire No. 20, same as No. 15	**$425-480.**
Empire No. 29, Bronze Top & Base, etc.	**$580-650.**
Cupid Empire, Bronze Top Nouveau Finish, 15-day, etc.	**$1,300-1,600.**

(More than 65 Empire styles were made)

Banjos:

Delaware, Key Wound, Brass Side Ornaments, c. 1900s	**$165-190.**
Mansfield, Electric-Time, Brass Side Ornaments	**$270-295.**
Brookfield, Electric-Time, Hour, ½-Hour, S.	**$185-245.**

*(Probably others. "New" Banjos are being
sold as original Willards!)*

Hall (All are 8-day, Weight, Striking Hour & ½-Hour on Cathedral Bell. The more expensive chime Westminster at the quarter hours on 4 Cathedral Bells & show Moon Changes):

No. 22, mahogany, old oak, oak natural, Chime & Moon	**$4,950-5,785.**
No. 2272C, mahogany, old oak, oak natural, Strike & Moon	**$3,400-3,800.**
No. 23, mahogany, golden oak, Strike	**$3,200-3,700.**
No. 24, mahogany, old oak, Weight, Strike, Chime & Moon	**$5,125-5,475.**
No. 25, same as above	**$6,200-6,775.**
No. 26, same as above	**$5,400-5,780.**
No. 27, golden oak, old oak, etc., Strike & Moon	**$5,900-6,875.**
No. 2774D, same as above, etc.	**$6,350-6,995.**
No. 2784, mahogany, golden oak, old oak, Cathedral Chime & Moon	**$7,200-7,870.**
Nos. 28; 2874D; 33; 3374D; 34; 3434D; 35; 3574D; 36; 3684D; 3684; 37; 3774D; 3784; 38; 39; 3974D. These clocks range in price	**$5,200-7,450.**

Wood Case, Black Adamantine Finish

(All have 8-day, ½-Hour Strike and/or Hour Strike, Cathedral Bell; the more expensive have Gold Plated Columns & Ornaments & Adamantine onyx Columns):

Lyons; Hull; Keswick	each **$145-175.**
Elba; Arno; Windsor; Sussex; Toulan; Pasha; Pequod; Domino; Don; Adnaw; Hastings; Pelham	each **$125-180.**
Texel; Bosnia; Tyro; Ideal; Wanda; Niphon; Ravenna; Manchester; Harrison; Petrol; Viking; Sparta; Delos; Milo	each **$160-195.**

Wood Case, White Adamanite Finish

(Works same as above):

Trent; Durban; Roslyn; Peru; Sucile; Adele; Chandos; Dalny; Sheffield; Leeds; Dale	each **$160-185.**

Metal Novelties, 1-day, Rich Gold Finish:

(More expensive models have Art Nouveau and/or Bronze Top & Base)

Ada; Bona; Cherubs; Corinna; Dido; Bungalow; Grapple	each **$68-78.**

Colonial; Cyril; Floss; Beth; Cis; Dimple; Vera; Ivan;
 Quaint; Nan   each **$78-86.**
Wagner, Shakespeare; Mozart; Schiller   each **$98-115.**
Vanity; Alice; Jess; Crispin; Joe; Piper; Serenade   each **$92-120.**

One-Day Novelties, Rich Gold and/or Bronze Art Finish:
Colin; Dorrit; Tick Tock; Paddock; Elephant; Natty;
 Holly; School Days; Eagle; Fountain   each **$72-92.**

WATERBURY CLOCK CO., Waterbury, Connecticut. Founded in 1857. Originally a part of Benedict & Burnham. Became a part of U. S. Time Corp. in 1944.

Regulators:

No. 3, walnut or mahogany, 8-day, Wgt., Time	**$1,450-1,700.**
No. 4, same as above except brass wgts.	**$1,600-1,850.**
No. 6, walnut, Standing Case, Brass wgts., 8-day, T.	**$2,275-2,450.**
No. 7, walnut, cherry, Cabinet Finish, glass sides, porc. dial, Brass wgts., 8-day, Time	**$3,800-4,250.**
No. 14, walnut, Cabinet Finish, Brass wgts., 8-d., T.	**$1,900-2,265.**
No. 61, quartered oak, walnut, 8-d., Brass wgts., T.	**$3,900-4,250.**
No. 71, quartered oak or mahogany, as above	**$3,900-4,475.**
No. 66, quartered oak, 8-d., T.	**$1,400-1,750.**

Wall:

General, white enamel, gilt molding, 30-d., Second Hand	**$385-425.**
12″ Drop Octagon, oak, 8-d., T.	**$245-280.**
Same, except 8-d., ½-Hour, S.	**$285-350.**
Same, except 8-d., T., Calendar	**$325-370.**
Same, except 8-d., ½-Hour, S., Cal.	**$360-395.**
12″ Arion, oak or mahogany finish, 8-d., T.	**$260-275.**
Same except 8-day, ½-Hour, S.	**$320-360.**
Same except 8-d., ½-Hour, S., Cal.	**$375-410.**
Same except 8-d., Y., Cal.	**$340-389.**
Yarmouth, oak or mahogany, 8-d., Spring, T.	**$550-620.**
Same except 8-d., Spring, ½-Hour Gong Strike	**$570-625.**
Regent, Rosewood Veneered, 8-d., T.	**$330-350.**
Same except 8-d., ½-Hr. S.	**$375-395.**
Same except 8-d., ½-Hr. Gong S.	**$395-420.**
12″ Heron, oak or mahogany finish, 8-d., T.	**$320-350.**
Same except 8-d., ½-Hr. S.	**$365-385.**
Same except 8-d., Cal.	**$400-425.**
Same except 8-d., ½-Hr. S., Cal.	**$450-485.**
Consort, oak or walnut, 8-d., T.	**$360-385.**
Same except 8-d., ½-Hr. S.	**$420-492.**
Same except 8-d., ½-Hr. Gong S.	**$400-453.**
Asbury, walnut, Cabinet Finish, 1-d., Wgt., S; also with alarm	**$522-586.**
Antique Drop, antique oak, 8-d., ½-Hr. S., Cal.	**$350-410.**
Carlton, walnut, 8-d., T.; 8-d., ½-Hr. Slow S.	**$785-865.**
Pontiac, same as above	**$850-1,100.**

Octagon Lever, R.C., rosewood or mahogany, 1-d., 4″, 6″, Spring	$140-170.
Same except 1-d., 8″, 10″, Spring	$160-180.
Same except 8-d., 6″, 8″, Spring	$192-222.
Same except 8-d., 10″, 12″, Spring	$220-260.
Drop Octagon, R.C., rosewood or zebra, Cal., 8-d., T., ½-Hr. S.	$340-375.
Same except Gilt finish	$245-270.
Same except Gilt finish & Calendar	$340-370.
Prescott, walnut, 8-d., T.; 8-d., ½-Hr. Slow S.	$520-560.
Ottawa, same as above	$560-585.
Breton, oak or mahogany, glass sides, 8-d., Spring, T.	$625-695.
Same except Gong S. added	$680-720.
Michigan, oak or walnut, 8-d., ½-Hr. S.; also Gong S.	$350-370.
Saranac, same as above except with barometer & thermometer	$360-410.
Waltham, Elgin, Bruce, oak, walnut or cherry, same works as above	each $652-687.
Study No. 4, oak, Cabinet Finish, glass sides, 8-d., Wgt., Gong S.	$860-985.
Augusta, oak, Cabinet Finish, Gilt Ornaments, 8-d., Wgt., Gong S.	$1,785-2,400.
Eton, oak or walnut, 8-d., Spring, T.	$650-780.
Same as Eton except Gong S.	$690-750.
Nassau, oak, 8-d., Sprg., ½-Hr. Gong S.	$680-725.
Nassau, oak, 30-d., Sprg., T., Second Hand	$800-925.
Walton, oak, 8-d., Sprg., T.	$470-575.
Same except 8-d., Sprg., ½-Hr. Gong S.	$510-625.
Same except 8-day, Sprg., T & Cal.	$650-750.
Same except 8-d., Sprg., ½-Hr. Gong S. & Cal.	$750-875.
Cairo, oak, walnut or mahogany, 8-d., Sprg., T	$625-825.
Same except 8-d., ½-Hr. Gong S	$700-925.
Calendar No. 33, oak or walnut, 8-d. T., Sprg., ½-Hr. Gong S.	$1,700-1,950.
Berlin, walnut, 8-d., wgt., T	$1,275-1,450
Toronto, walnut, 8-d., Sprg., S.	$950-1,200.
Same except 8-d., Sprg., T & S	$1,100-1,400
Cambridge, oak or walnut, 8-d., Sprg., T	$850-975.
Same except 8-d., Sprg., ½-Hr. Gong S	$850-1,100.
Same except 30-d., Sprg., Time with Second Hand	$995-1,450.
Study No. 3, oak, 8-d., wgt., ½-Hr. Gong S	$700-850.
Dresden, oak, 8-d., Spring, ½-Hr. Gong S.	$925-1,250.
Same except 30-d., Sprg., T & Second Hand	$1,200-1,475.
Pictou, oak or mahogany, glass sides, 8-d., Sprg., T	$875-985.
Same except 8-d., Sprg., ½-Hr. Gong S	$900-1,200.
Nelson, oak or mahogany, glass sides, 8-d., Sprg., T	$850-995.
Same except 8-d., Sprg., ½-Hr. Gong S	$1,000-1,450.
Calendar, No. 25, quartered oak, or mahogany, 8-d., Wgt., Y	$2,450-2,775.

Crystal Regulators (All are Cast Gilt bezel, beveled glass front, sides & back;

Ivory dial, Visible Escapement, mercury pendulum, 8-day, ½-Hour Gong Strike. Also, Rich Gold Plated):

Mogul	**$750-900.**
Flanders, Vendee	**$450-675.**
Rennes, Bordeaux	**$785-965.**
Brest, Vannes	**$325-465.**
Gers, Girande, Gard, Green Onyx Base & Top	**$875-990.**
Savoy, Cantel, Dieppe	**$375-455.**
Brittany	**$525-665.**
Ostend	**$325-399.**

Shelf (All walnut, 8-d., ½-Hr. S.; 8-d., ½-Hr. Slow S. and/or Gong):

LaMar; Borden; Morris; Sussex; Dayton; Merwin; Arcade; Kimble	each **$220-270.**
Essex; Middlesex, ash with walnut trimmings, 8-d., ½-Hr. S.	each **$170-198.**

WELCH, SPRING & CO., Forestville, Connecticut. Founded by Elish N. Welch and Soloman C. Spring, 1868. Merged with E. N. Welch Mfg. in 1884.

Calendars:

Wagner, Mantel B.W., 8-d., T & S, Spring, V Calendar Mechanism	**$1,780-2,350.**
Auber, Mantel B.W., 8-d., T & S, Wgt., Strap Brass Movement, V Cal.	**$2,700-3,200.**
Italian-type, shelf, rosewood, 8-day, Spring, V Cal. movement	**$1,200-1,665.**
Regulator No. 11, rosewood, 8-d., TS, Spring, V Cal. movement	**$2,800-3,460.**
Wagner, Hanging B.W. walnut, 8-d., TS, Spring, V Cal. movement	**$3,400-3,995.**
Gale, wall, walnut, 8-d., pendulum	**$4,450-4,785.**

E. N. WELCH MFG. CO., Bristol (Forestville), Connecticut. Formed with a merger of Welch, Spring & Co., in 1884. Stayed in business, and in 1903 the firm's name was changed to Sessions Clock Co. (see).

Calendars:

Drop Octagon R.C., rosewood, 8-d., TS, Simple Calendar	**$385-600.**
Victorian Kitchen, shelf, walnut, 8-d., TS	**$800-975.**
Eclipse Regulator, walnut, 8-d., T, Spring, Simple Cal. (made by Welch for Metropolitan Mfg. Co., New York)	**$850-1,100.**
Eclipse Regulator, oak, 8-d., T, Spring, Simple Calendar	**$375-475.**

Alarms:

Fairy Queen, 8-d., Lever, Calendar	**$82-98.**
Same except 8-d., Lever only	**$42-60.**
Fire Bug, 1-d., Lever, nickel	**$110-125.**
Nelson, 1-d., nickel	**$42-60.**

Daybreak, 1-d., Lever   **$42-60.**

Shelfs:

Assortments A thru K, Nos. 1 thru 65, 8-d., ½-Hr.
    S., Wire or C. Bell, with or without bell, walnut   each **$150-180.**

Dewey; Sampson; Schley; The Maine; Lee; Wheeler;
    oak, same movements as above   each **$270-295.**

Black Enameled Wood, Mantels:

Leno, 8-d., C. Bell, American Sash, White, Gilt or
    Fancy Perforated dials   **$92-120.**

Yebba, same as above   **$115-145.**

Zella, same as above   **$140-165.**

Ulmar, Shafter; same as above   **$142-162.**

Roosevelt; Belasco; Sorma; Nansen; Stagno; Alberta;
    Andree; De Merode; Burkhart; De Reszke; Calve;
    Karina; same movements as above   each **$165-185.**

Viarda, Nos. 1, 2, 3, same movements   each **$163-172.**

Pinero, Nos. 1, 2, 3, same movements   each **$168-175.**

Nicolini, A, B, C, same movements   each **$155-167.**

Ursula, Nos. 1, 2, 3, same movements   each **$163-174.**

Regulators:

A, black walnut, mahogany or antique oak, 8-d.,
    1 wgt., 3-mercury pendulum   **$3,700-4,200.**

C, same as above   **$3,500-3,750.**

E, walnut, ash or mahogany, Hanging, 8-d., Wgt., T   **$1,675-1,895.**

F, same as above   **$1,650-1,875.**

G, same as above   **$1,400-1,700.**

H, same as above except 8-d., Spring, T.   **$525-625.**

H, same as above, except 8-d., Spg., TS   **$575-655.**

H, same as above except 8-d., Spg., Cath. Bell   **$675-785.**

No. 7, polished black walnut, 8-d., T., Sweep Second   **$2,250-2,550.**

No. 8, same as above   **$1,750-1,975.**

No. 11, same as above   **$1,400-1,700.**

No. 12, same as above   **$1,600-1,895.**

Clock Museums, United States and Canada

California:

California Academy of
Sciences, Science Museum
Golden Gate Park
San Francisco

Colorado:

Hagans Clock Manor Museum
Highways 68 & 74
Evergreen

Connecticut:

P.T. Barnum Museum,
804 Main Street
Bridgeport

Wethersfield Historical Society
150 Main Street
Wethersfield

Winchester Historical Society
225 Prospect Street
Winsted

District of Columbia:

United States Naval Observatory
Massachusetts Avenue at 34th Street N.W.
Washington

Illinois:

Illinois State Museum of
Natural History & Art
Spring & Edwards Streets
Springfield

Indiana:

Bartholomew County Historical Society

Court House
Washington Street
Columbus

Union County Historical Society
26 West Union
Liberty

Iowa:

The Bily Clock Exhibit
Horology Museum
Spillville

Kansas:

Harbaugh Museum
National Bank of Commerce Building
Wellington

Wichita Historical Museum
3751 East Douglas Avenue
Wichita

Maine:

Parson Fisher House
Blue Hill

Maryland:

Maryland Historical Society
210 West Monument Street
Baltimore

Massachusetts:

The Averys Antique Clocks Collection
219 Lincoln Avenue
Amherst

Essex Institute
132 Essex Street
Salem

Old Sturbridge Village
Sturbridge

Michigan:

Henry Ford Museum and
Greenfield Village
Dearborn

Mississippi:

The Old Place
Highway 90
Gautier

New Hampshire:

Million Dollar Schuller Museum
Farmington

The Clock Museum
43 Park Street
Newport

New York:

Old Red Barn Museum
Salem-Greenwich Road
Salem

Yorker Yankee Village
Ireland Road
Watkins Glen

Ohio:

Knox County Historical Society
Newark Road
Mount Vernon

Licking County Historical Society
6th Street Park
Newark

Pennsylvania:

The Columbia Museum of
Horological Antiquities
333 N. 3rd Street
Columbia
(Note: Headquarters of the National Association
of Watch and Clock Collectors)

Hershey Museum
Park Avenue & Derry Road
Hershey

Texas:

Bosque Memorial Museum
Avenue Q
Clifton

Vermont:

Morristown Historical Museum
1 West High Street
Morrisville

Canada:

Huron County Pioneer Museum
Central Public School
North Street
Goderich, Ontario

Editor's note: Write to the museum of your
choice before visiting. Inquire about admission
fees, days and hours to visit.

Clock Terms and General Information

(with abbreviations often found in clock books)

"Accutron:" a trade name used by the Bulova Watch Co., New York City, 1960 and later, for their electronic battery-powered timepieces.

Acorn Clock: Called this because of its shape.

Adjusted: An adjusted pendulum clock is corrected only for temperature error. High grade watches have been also adjusted for positional errors. Some watches have the word "adjusted" engraved on their movements.

adv.: Advertised.

Alarm: Marking a predetermined time; found on early Tall Clocks as well as Shelf Clocks.

AM.: American.

American Clock Co., Bristol, Connecticut and New York City. From 1849 to 1879, distributor for all clocks made in Connecticut.

Ammonia Maser: The vibration of the Ammonia molecule is the timekeeping element. Not as accurate as the Atomic Clock. "Microwave Amplification by Simulated Emission of Radiation"—MASER.

Anchor Escapement: Used on most pendulum clocks today, it supplanted the verge escapement and made very accurate clocks possible.

Anniversary Clock: Another name for a Year Clock.

Antiquarian Horological Society: Address: 35 Northampton Square, London, E.C.I., England. Organized in 1953 to encourage research and to preserve examples of clocks, watches, etc. Serious horologists should join this fine society.

apt.: Apprentice.

Arabic Numerals: More convenient to read than Roman numerals, they were brought to Europe by the Crusaders in the twelfth century. Not commonly used on clocks until recent years.

Arbor: The axle or shaft upon which the gears or wheels of a clock are mounted.

Arc: The angle through which the balance wheel or pendulum swings.

Architectural Clock: The style of the case is based on the architectural lines of certain buildings.

Armillary Sphere: The emblem of the British Horological Institute (see), it's a fixed model of the universe, made up of circles or rings, in use for more than 2,000 years.

Artificial Clock: A mechanical clock; an early name to differentiate it from a natural "clock" (see).

astro.: Astronomical.

Astronaut's Clock: Unaffected by varying gravity, etc., an extremely accurate timekeeper developed for travel in space.

Astronomer's Pendulum: A weight which swung on a cord to measure time. The time of swing was calculated by Galileo, among others.

Atomic Clock: Because the radio frequencies emitted by atoms and molecules at low pressures are fixed and unchanged by time, they can be used to control a radio oscillator which controls a Quartz Clock. Using the vibration of the Caesium atom as its "pendulum" which is at 9,192,631,770 cycles per second.

Back Plate: Two plates hold the arbor of a clock train; the back plate is farthest from the dial.

Balance: The controller (or oscillating wheel) with the hairspring that regulates the escape of power

Balance Cock: Normally a shock absorber, it holds the bearing for one end of the Balance (see).

Balance Spring: Another name for the Hair Spring (see).

Balance Staff: The staff (arbor) upon which the balance is mounted.

Balloon Clock: Shaped like hot-air balloons of the late 18th century; these were bracket (table or shelf) clocks.

Banjo Clock: Simon Willard introduced these in 1802. Originally called Improved Timepieces, because of their shape they are called Banjos today.

Barrel: This round container holds the mainspring.

Basket Top: A popular form of top on English bracket clocks, late 17th century.

Baton: A stroke instead of a number on a clock dial.

B.D.: Black Dial.

B.D.V.E.: Black Dial, Visible Escapement.

Beat: The tick of the clock; the sound of the teeth of the escape wheel on the pallets. When the intervals of the tick are uneven, it's called "out of beat."

Bell Top: A popular form of top on English bracket clocks, late 17th century.

Bench Key: Sometimes with adjustable ends, a star-shaped key with ends to fit different winding squares.

Bezel: The metal rim which holds the glass over the clock's dial.

Bob: The wire loop on the pendulum which is threaded for the regulating nut.

Bracket Clock: Generally, all spring-driven clocks designed to stand on a table, bracket or shelf.

Braille Clock: Raised numerals on the dial enable a blind man to "feel" the time.

Bridge: Fixed at both ends, it's a metal bar which carries one or more pivot bearings.

British Summer Time: A Parliamentary Order (yearly) in the United Kingdom advances the Greenwich Mean Time by an hour each summer.

Bushing: The point where the arbor extends through the clock plate for its "bearing," when more metal is added to make the bearing surface longer than the original thickness of the plate.

C.: Clock or Clockmaker.

Cal.: Calendar.

Calendar Clock: A type of bracket clock made in large quantities after 1860; also, an attachment on tall clocks to tell the day of the month.

Cannon Pinion: The pinion which carries the minute hand.

Cannon Tubes: A misnomer applied to the hour pipe which carries the hour hand.

Carriage Clock: A small, portable clock, usually in a leather carrying case, spring-driven with balance, each has a platform escapement. Most were T and S, some had chimes, some were repeaters.

Cartel Clock: Of the Louis XV Period, an ornate French wall clock, in a carved wood or cast bronze case, with a gilt finish and half-seconds pendulum.

cent.: Century.

Center Seconds Hand: This type is mounted at the center of the dial; also "sweep seconds hand."

Chapter Ring: The part of the dial on which the hours are marked.

Chapters: The hour marks of a clock.

Chime: A simple melody on bells to denote the quarter or half and preceeding the hour. Notre Dame Cathedral in Paris is an eight-bell chime; the Whittington chime of Bow Church in London is eleven (can be modified); Westminster is a four-bell.

Chiming Clock: A clock that chimes.

Chronometer: A clock which has passed strict observatory tests; a very accurate timekeeper.

Clepsammia: An hour glass filled with sand; also called a "sand thief."

Clepsydra: An early form of timekeeper regulated by a flow of a liquid.

Click: The noise caused by the teeth of the ratchet-wheel escaping past the click during the winding; a spring-tensioned pawl holding up against the tension of the mainspring, enabling it to be wound.

Cock: Fixed at one end only, a metal bar which carries one or more pivot bearings.

Collet: A solid or split brass ring used for securing the minute hand or the inner end of the balance spring to the balance staff.

Compensation pendulum: The distance between the center of gravity of the pendulum and its anchorage remains constant at varying temperatures. The two principal forms are the Harrison pendulum and the Graham mercurial pendulum.

col. Collection.

Column Clock: Clock with a base like an architectural column.

Conical Pivot: Basic cone-shaped pivot used in cheap clocks, it runs in a cup bearing.

Controller: The balance or pendulum which controls the clock's timekeeping.

Conversion: Older timepieces were converted or "up-dated" as improvements were invented, such as the pendulum, lever escapement, etc. Sometimes it's difficult to tell the date of very old clocks because of this. **NEVER** convert an antique clock! For shame!

Cordless Clock: American name for a battery operated clock.

Count wheel: A locking plate that controls the number of blows struck on the bell or gong of a clock; it prevents repeat striking.

chron: Chronometer.

Crown Wheel: The escape wheel of a verge escapement. The Anchor Escapement's invention made it unnecessary in clocks. It was used for more than 400 years.

Crutch: The claw by which the pendulum is attached to the escapement, on an arm attached to the anchor shaft or arbor.

Cup Bearings: Associated with the conical pivots of the balances of cheap clocks, it's the cone-shaped depressions in the end of a steel screw or plug.

Cut Balance: A temperature compensation balance made of steel and brass strip joined together so they bend when heated to give temperature compensation.

D.: Dealer.

Dead Beat: The movement in definite jumps without recoiling.

Decimal Timer: Calibrated with 100 instead of 60 divisions to the minute, it's used for certain industrial purposes.

Detent: A lever that locks something or holds up a spring. **Detente** — "end of strained relations." Used specifically for the lever associated with the detent escapement.

Dial: The face of a clock.

Dial Plate: That to which the chapter ring and spandrel movements are attached.

dir.: Directory.

Domestic Clock: Any type used in the home.

Drum: In a weight-driven clock, the round barrel on which the weight gut or cord is wound.

Dumb Repeater: Where the hammer strikes a fixed metal block instead of a bell or gong.

Dutch Wag: "wag on the wall" clock; the short pendulum wags rapidly beneath it.

Eight-Day Clock: Has to be wound only once a week. The extra day is for reserve.

engr.: Engraved.

Equation Clock: A type of clock that shows the difference between solar time and mean time.

Escapement: The method of regulating the release of power, assuring the accurate recording of the passage of time.

Escape Wheel: The last wheel in a going train, it's controlled by the escapement.

False Bob: Another name for a mock pendulum.

Finial: Acorns, flambeaux (flames), spires, etc.—all used to decorate the top of a clock case.

Fly: Rotating fan (or governor) used to slow down a chiming or striking mechanism.

Fourth Wheel: Usually the wheel which carries the second hand and drives the escape wheel; it is the fourth wheel from the great wheel in the going train of a clock.

Frame: The structure in which the parts of a clock are built.

Free Pendulum Observatory Clock: Accurate to within 1 second per year.

F.S.: French Sash.

Friction Roller: As a bearing for pivots, it needs no lubrication.

Friction Spring: A clutch for hand setting; also for taking up backlash of a center second hand.

Front Plate: Two plates hold the arbor of a clock train; the front plate is nearest to the dial.

Front Wind: Clock wound through the dial.

Fuzee: A cone-shaped piece with a spiral track cut around it; mounted on the great wheel arbor, it acts as a "gear changer." As the spring runs down, it preserves, more or less, a constant torque on the train as long as the clock is running.

Gadget Clock: A clock with extra accessories such as moving figures.

Gimbal: A support used to keep a clock level.

G.F.: Grandfather clock.

G.M.: Grandmother clock.

Gnomon: Also called a "style"; the pin, rod or wedge-shaped plate of a sun-dial that throws the shadow on the dial.

Going Barrel: Contains the mainspring; the most common spring motor for clocks.

Going Train: Responsible for its timekeeping and controlled by the escapement, the train of gears in a timekeeper.

Grandfather Closk: The popular name for a long-case clock.

Grandmother Clock: The popular name for a miniature long-case clock.

Gravity Arm: Pivoted at one end, a small lever of a certain weight for giving impulses to a pendulum.

Great Wheel: The first wheel in the train. On the drum, it is weight-driven; on the going barrel or on the fusee, it is spring-driven, depending upon which type is used.

Hair Spring: Balance spring of a clock.

Hands: Used to mark hours, minutes, or seconds on a clock dial.

Hanging Clock: A weight-driven clock that is hung from the wall.

Hooke's Law: The force produced by a spring is proportional to its tension.

Horology: The science of measuring time, or the principles and art of constructing instruments for measuring and indicating portions of time.

ill.: Illustration.

imp.: Importer.

Impulse and Locking: Dead-beat escapements have two actions: impulse is the period during which the train imparts impulse to the balance or pendulum; during the rest of the time the train is locked.

insts: Instruments.

Inverted Bell: A form of top to an English bracket clock, 18th century.

Iron-front Clock: Connecticut Shelf Clock with cast-iron front.

Isochronism: Occupying equal time.

Jack: A moving figure turned by a clock mechanism.

Jesuit Clocks: Jesuit missionaries took this style clock to China after 1585.

J.: Jeweler.

Jewel: Synthetic or semiprecious stones used for bearings.

Kitchen Clock: Just that, a wall clock for kitchen use.

Largest Clock Dial: 50 feet across, it's on the Colgate-Palmolive plant in Jersey City, N.J. Built in 1924, the minute hand is 27 feet 3 inches long.

Leaves: The teeth of the pinion gears.

M.C.: Marble Column.

Mainspring: The spring which supplies driving power in a spring-driven clock.

Main Wheel: The driving first wheel, usually attached to, or part of, the barrel.

Marine Chrononometer: Accurate to within 0.3 seconds a day.

Mean Time: Where all days and hours are of equal length. This is opposed to Solar Time where all days are not of equal length.

mfg.: Manufactured.

mfr.: Manufacturer.

Millisecond: A thousandth of a second.

min.: Minute.

Minute Hand: Hand on clock which indicates minutes.

Mock Pendulum: A disc fixed to the Pallets swings to and fro in a curved slot near the top of the dial; to show the clock is going. Also called a "false bob" (see).

Monumental Clock: One built like a monument.

Most Accurate (and complicated) Clock: The Olsen Clock, Copenhagen (Denmark) Town Hall. Has more than 14,000 units; its mechanism functions in 570,000 different ways.

Motion Work: Twelve to one gearing driving the hour hand from the minute hand.

mov't.: Movement. "The works" of a clock.

mus.: Musical.

Musical Alarm: Instead of ringing a bell a tune is played on a small music box. Still being made, originally produced in large numbers by the Germans and the Swiss, late 1800s to World War I.

n/d: No date.

n/p: No place.

Nanosecond: thousand-millionth of a second — for measuring electronic time intervals.

Natural "Clock": The sun or moon.

Numerals: Hour numbers on a dial. Usually Arabic; once almost universally Roman.

OG or ogee: A wave-like molding, one side convex, the other concave, shaped like the letter "S." From 1825 to 1915, six or more sizes were used. Jerome is credited with reviving the clock industry after the 1837 Depression by using this style of clock case.

Oil Sink: For retaining oil; it's a small hollow cut into the outside of a plate of a clock, concentric with a pivot-hole.

Oldest Mechanical Clock: Completed in China in 725 A.D. by I'Hsing and Liang Ling-tsau.

Orrery: A machine in a planetarium that shows the relative position of the planets.

Out of Beat: If the action of the pendulum releasing a tooth of the escape wheel at the end of each swing is not symmetrical, the clock is "out of beat."

Overcoil: A hairspring with the outer quarter turn raised and curved toward the center thus avoiding becoming lopsided as it opens and closes.

P & S: Pillar and scroll clocks.

Pallet: The parts of an escapement which intercepts the teeth of the escape wheel and receives an impulse.

pat.: Patent.

Pediment Top: A style of top in the form of a triangular pediment.

pend.: Pendulum.

Pendulum: Intercepts the teeth of the escape wheel and receives an impulse.

pat.: Patent.

pend.: Pendulum.

Pendulum: A weight swinging under the influence of gravity. The pendulum controls the rate of a clock movement and is kept swinging from impulses from the escapement.

Pendulum Rod: It holds the bob, made of steel, brass, etc.

Pillar and Scroll Clock: A shelf clock attributed to Eli Terry.

Pillars: Rods, riveted or screwed, which fasten the plates together to make a clock frame.

Pinion: A small toothed wheel, usually driven by a gear. Six teeth, or leaves, are usual—more teeth in higher-quality clocks.

Pin-wheel Escapement: Where the locking and impulse is effected by pins mounted near the rim of the escape wheel, at right angles to the plane of the wheel.

Plates: The front and back of the clock movement.

Pivot: The small end or an arbor or shaft that runs in a bearing hole.

Position Regulator: A regulator.

Point of Attachment: Usually, the point at which the inner end of the hairspring is fixed to the balance.

Potence: A form of cock (see) used as a lower bearing.

Prinked: Small dots impressed in rows on a clock plate to harden the surface.

Q.: Quarter.

Quarter Chime: More than two bells strike at each quarter.

Quarter Repeater: A repeater which records the hours and quarter hours.

Quail Clock: Operates like a cuckoo clock but its bellows imitate a quail.

Quartz Clock: Accurate to within 1 second in 30 years.

Rack and Snail: The rack is a bar with teeth on one edge; the snail is a snail-shaped cam. Simply, a striking mechanism which allows the strike to be repeated.

Rate: A "good rate" is a good timekeeper; if it gains or loses, it is said to have a gaining or losing rate.

Rating Nut: Used to alter the clock's rate, it's located below the pendulum bob.

Recoil: Backward rotation of the escape wheel in a recoil escapement.

Recoil Escapement: An escapement in which recoil takes place.

Regulation: Adjusting the rate of a clock.

Regulator Clock: Any accurate wall clock in the early days; later the term was applied to many Connecticut wall clocks.

Republican Time: The attempted decimal division of the hour into 100 minutes by the French Revolutionists in 1793.

Repeater: Generally a clock which strikes the last hour and the divisions of an hour only when the striking train is set in motion by moving a lever or pulling a cord.

rep.: Repeating.

Riefler Clock: Accurate to within 1/100 of a second per day.

Second: The International Committee of Weights and Measures defines the second as one 31,556,925-9747th part of the year 1900.

Seconds Hand: Hand of a clock showing seconds.

Seconds Pendulum: A pendulum slightly more than 39 inches long, beating seconds.

Set Hands: Every timepiece has a clutch mechanism that allows the hands to be set to a new time without damaging the movement.

Set-Up: When the clock is fully run down, the residual tension to which a mainspring is set.

sd.: Signed.

Ship's Time:

No. of Bells	Hour (A.M. or P.M.)		
1	12:30	4:30	8:30
2	1:00	5:00	9:00
3	1:30	5:30	9:30
4	2:00	6:00	10:00
5	2:30	6:30	10:30
6	3:00	7:00	11:00
7	3:30	7:30	11:30
8	4:00	8:00	12:00

sig.: Signature.

spg.: Spring.

Spring Barrel: The barrel containing the mainspring.

Spring Clock: Springs rather than weights are the motive power.

Stop Work: A device which prevents overwinding.

str.: Striking.

Striking Clock: These sound the hours on a bell or gong.

Style: See Gnomon.

Supplementary Arc: The portion of the arc of a pendulum after impulse has ceased.

Suspension: Flexible mounting of a pendulum, usually a suspension spring (see).

Suspension Spring: The pendulum is suspended on this spring.

Sweep Seconds: Mounted in the center of the clock dial, this hand sweeps the full area of the dial.

7Tall Clock: Long-case, floor or hall clock, nicknamed Grandfather.

Tavern Clock: Also generally known as an Act of Parliament Clock.

T. & S.: Time and strike. Clock tells the time and strikes on the quarter, half, three quarter and/or hour.

T. only: Clock tells the time but has no strike.

Tell-Tale Clock: Another name for a watchman's clock.

tp.: A clock that doesn't strike.

T.S. & A.: Time, Strike and Alarm. Clock tells the time, strikes and also has an alarm bell.

Thirty-Hour Clock: Has to be wound daily and runs at least thirty hours.

Tick: The sound made by the release and arrest of a train of gears by the escapement. The sound of a "tick" lasts about 15 milliseconds.

Tic-Tac Escapement: An anchor escapement in which the anchor embraces only one and a half teeth.

Timing Washers: Placed under the screws in the rim of a balance to adjust the rate at which it swings; also to correct poise errors.

Time and Strike: Any clock that strikes as well as telling the time.

Timepiece: Any clock that does not strike or chime.

Tock: The partner of "tick" (see). Ticks differ from "tocks" because alternate pallets arrest the escape wheel and different parts of the movement resonate. The sound of a "tock" lasts the same time as a "tick."

Tower or Turret Clock: A steeple, church or public clock in a tower.

Train: The wheels and pinions of a clock.

V.E.: Visible Escapement.

Verge: The pallet axis of a clock.

Verge Staff: The arbor upon which the pendulum, crutch or balance is mounted is the verge or verge staff.

V.P.: Visible Pendulum.

W.D.: White Dial.

W.D.V.E.: White Dial, Visible Escapement.

Wagon Spring: Joseph Ives invented this type of flat-leaved spring used to power a clock movement.

Wall Clock: Any type of clock that hangs on the wall.

Westminster Chime: Chime of Westminster Palace Clock ("Big Ben") which comes from the fifth bar of Handel's *Messiah*—"I know that my Redeemer liveth." Originally used at St. Mary's church, Cambridge, England, 1793-1794.

wgt.: Weight.

Index